Operator Theory
Advances and Applications
Vol. 99

Editor:
I. Gohberg

Measures of Noncompactness in Metric Fixed Point Theory

J.M. Ayerbe Toledano
T. Domínguez Benavides
G. López Acedo

Springer Basel AG

Departamento de Análisis Matemático
Facultad de Matemáticas
Universidad de Sevilla
41080 Sevilla
Spain

1991 Mathematics Subject Classification 54H25, 46A50, 55M20, 47H10

A CIP catalogue record for this book is available from the
Library of Congress, Washington D.C., USA

Deutsche Bibliothek Cataloging-in-Publication Data
Ayerbe Toledano, J. M.:
Measures of noncompactness in metric fixed point theory / J. M. Ayerbe Toledano ;
T. Domínguez Benavides ; G. López Acedo. – Basel ; Boston ; Berlin :
Birkhäuser, 1997
  (Operator theory ; Vol. 99)
  ISBN 978-3-0348-9827-0   ISBN 978-3-0348-8920-9 (eBook)
  DOI 10.1007/978-3-0348-8920-9

© 1997 Springer Basel AG
Originally published by Birkhäuser Verlag in 1997
Softcover reprint of the hardcover 1st edition 1997

Printed on acid-free paper produced from chlorine-free pulp. TCF ∞
Cover design: Heinz Hiltbrunner, Basel

ISBN 978-3-0348-9827-0

9 8 7 6 5 4 3 2 1

Contents

Preface

What is clear and easy to grasp attracts us; complications deter

David Hilbert

The material presented in this volume is based on discussions conducted in periodically held seminars by the Nonlinear Functional Analysis research group of the University of Seville.

This book is mainly addressed to those working or aspiring to work in the field of measures of noncompactness and metric fixed point theory. Special emphasis is made on the results in metric fixed point theory which were derived from geometric coefficients defined by means of measures of noncompactness and on the relationships between nonlinear operators which are contractive for different measures.

Several topics in these notes can be found either in texts on measures of noncompactness (see [AKPRS], [BG]) or in books on metric fixed point theory (see [GK1], [Sm], [Z]). Many other topics have come from papers where the authors of this volume have published the results of their research over the last ten years. However, as in any work of this type, an effort has been made to revise many proofs and to place many others in a correct setting.

Our research was made possible by partial support of the D.G.I.C.Y.T. and the Junta de Andalucía.

Many thanks are due to the readers of different parts of the manuscript, in particular to our friend Brailey Sims who made a careful reading of the original version, to our friend Stanisław Prus who suggested proofs of several results and to the members of our research group. We are deeply indebted to Lesley Burridge for careful revision of the language style and to our friend Juan Arias de Reyna for his help with TEX and his fruitful mathematical remarks.

Seville, December 1996

José María Ayerbe Toledano
Tomás Domínguez Benavides
Genaro López Acedo

Introduction

By a metric fixed point theorem we mean an existence result for a fixed point of a mapping f under conditions which depend on a metric d, and which are not invariant when we replace d by an equivalent metric. The best known metric fixed point theorem is the Banach theorem, also called the contractive mapping principle: "Every contraction from a complete metric space into itself has a (unique) fixed point". It is clear that a contractive mapping can lose this property if d is replaced by an equivalent metric. The Banach theorem is a basic tool in functional analysis, nonlinear analysis and differential equations. If we relax the contractive condition, requiring only that the mapping be nonexpansive, that is, $d(f(x), f(y)) \leq d(x, y)$, then trivial examples show that the Banach theorem need no longer hold. This failure may have been the reason why no significant result about the existence of fixed points for nonexpansive mappings was obtained for many years. However, in 1965, Browder [Br1 and Br2], Göhde [Go] and Kirk [Ki1] proved the following results: "Let X be a Banach space, C a closed, convex and bounded subset of X and $T : C \rightarrow C$ a nonexpansive mapping. If X is either a Hilbert space, or a uniformly convex Banach space or a reflexive Banach space with normal structure, then T has a fixed point". This result is, in some sense, surprising because it uses convexity hypotheses (more usual in topological fixed point theory) and geometric properties of the Banach spaces (commonly used in linear functional analysis, but rarely considered in nonlinear analysis prior to this time). The above results were the starting point for a new mathematical field: the application of the geometric theory of Banach spaces to fixed point theory. The texts [GK1], [AK], [KZ] and [Z] constitute excellent surveys of this theory.

The Brouwer [Br, 1912] fixed point theorem should be considered in a different setting: "Every continuous mapping from the unit ball of $\mathbb{R}^n$ into itself has a fixed point". A generalization of Brouwer's theorem was obtained by Schauder [S, 1930]: "Every continuous and compact mapping from a closed, convex and bounded subset C of a Banach space X into C has a fixed point". Clearly the conditions in the hypotheses are preserved if the norm of X is replaced by an equivalent norm, so this theorem cannot be viewed as a metric fixed point theorem. The situation is completely different when certain generalizations are considered, in particular those concerning ϕ-contractive or condensing mappings. A ϕ-contractive or condensing operator is a mapping under which the image of any set is, in a certain sense, more compact than the set itself. The degree of noncompactness of a set is measured utilizing functions ϕ called measures of noncompactness. The first such measure was defined by Kuratowski [Ku1, 1930]. Later other measures were defined by several authors (for instance [GGM, 1957] and [I1, 1972]). Measures of noncompactness have proved to be a very useful tool in the theory of functional equations, including ordinary differential equations, partial differential equations, integral and integro-differential equations, optimal control theory, etc. Kuratowski's initial interest in these measures was connected with certain problems in general topology, but the concept received a new impetus after the work of Darbo [D, 1955] which generalized the Schauder fixed point theorem to the class of set-contractive operators, that is, operators which satisfy $\alpha(T(A)) \leq k\alpha(A)$ with $k < 1$, $\alpha(\cdot)$ being the Kuratowski measure of noncompactness. After this result several fixed point theorems for set-contractive or condensing operators were proved (see for example [Sa1, 1967] or [Pe, 1972]). The most significant difference between these results and Schauder's theorem (besides greater generality) is their metric character. Again, the conditions in the hypotheses are not preserved under renorming. On the other hand, though all the usual measures of noncompactness are equivalent in a topological sense, the ϕ-contractiveness constant of the mapping is not preserved when different measures are considered. Moreover, the geometric characteristic of the Banach space plays a central role when comparing the ϕ-contractiveness of an operator for different measures of noncompactness.

In recent years, measures of noncompactness have also proved to be very useful in metric fixed point theory (see for instance [GK1], [Ba2], [DL1], [DL2], [Do6], [DX], [Pr2], [Zh1], [Zh2]), yielding existence or stability results for nonexpansive and uniformly Lipschitzian mappings based upon certain coefficients defined in terms of such measures.

The goal of this book is to develop, in a self-contained form, those results in metric fixed point theory which involve the use of measures of noncompactness. Although some of these results have already been considered in [GK1], most results in this book have not been included in any previous text book. To place them in the correct setting it is necessary to also study several geometric properties of Banach spaces. For the sake of simplicity, we will always consider real Banach spaces. Special emphasis is given to the study of the effect of the geometry of the space on the relationship between ϕ-contractions for different measures of noncompactness. Most results of this kind were obtained by the authors over the last ten years and have not previously appeared in any textbook.

The book is organized in thirty five sections arranged in ten groups called chapters. Every chapter has an introduction which explains what you will find there and how it is related to earlier chapters. References are indicated by letters (followed by numbers if necessary) in square brackets. The bibliography contains most of the relevant books about this subject, but the selection of research papers does not pretend to be exhaustive. We have also included a subject index and a list of symbols and notations in order to make the reading easier. The numbering of the results is standard: For example, Theorem II.3.4 means Theorem 4 of the Section 3 in Chapter II and Example II.2 means Example 2 in Chapter II. This book is mainly addressed to postgraduate students who want to learn something about metric fixed point theory and to researchers in the area. We assume that the reader is familiar with the basic results of real analysis, functional analysis and Banach space geometry. An acquaintance with the main results in [B], [Da1] and [Di] certainly suffices. Occasionally deeper results (Dvoretsky's theorem, Milman's characterization of nonreflexivity, convexity inequalities, etc.) are needed. Precise references are given in these cases.

Chapter I is devoted to the Brouwer and Schauder theorems. It provides basic material for proving fixed point theorems concerning measures of noncompactness. In contrast with the contractive mapping principle, the proof of the Schauder theorem depends on a profound topological result, namely, that there is no continuously differentiable map from a closed ball in n-space into its boundary which leaves the boundary pointwise fixed. An approximation process then implies the Schauder fixed point theorem. Some applications of these theorems to the study of differential equations have been included.

In Chapter II we study different measures of noncompactness. An axiomatic approach to this notion has been adopted with the axioms selected so that useful conclusions follow naturally from them. We also prove the essential properties of the three most usual measures (Hausdorff measure, Kuratowski measure and separation measure). The final part of the chapter is devoted to fixed point theorems for condensing operators and some applications to differential equations.

In Chapter III the notion of minimal set with respect to a measure of noncompactness is introduced. This concept appeared in [Do1, 1986] and has proved to be a very useful tool for obtaining relationships between ϕ-contractions for different measures of noncompactness and for simplifying proofs in which measures of noncompactness are involved. Spaces where we can obtain minimal sets without decreasing the measure have special importance. Such spaces, called minimalizable spaces, are very significant for later work.

In Chapter IV we consider geometric notions such as uniform convexity and uniform smoothness which play an important role in metric fixed point theory. Special attention is given to two dual notions: k-uniform convexity and k-uniform smoothness. These notions have been rarely considered in literature on this topic. This chapter may be seen as a bridge between the usual geometric properties considered in metric fixed point theory and geometric properties involving ideas of noncompactness, whose importance in fixed point theory is emphasized in Chapter VI.

Chapter V is devoted to certain geometric properties of Banach spaces which are defined by mean of measures of noncompactness, or which are related to them. The notion of near uniform convexity is studied and the corresponding modulus is introduced and, in some cases, computed. As a dual notion we study nearly uniformly smooth spaces. The uniform Opial condition is also considered. Some connections between this property and special values of the modulus of near uniform convexity for the Hausdorff measure of noncompactness are shown.

  In Chapter VI we apply the geometric and compactness conditions included in Chapters IV and V to metric fixed point theory. In order to assure normal structure and stability of the fixed point property, we introduce Bynum's coefficients of normal structure [By3, 1980]. Those can be understood as a measure of the normal structure of Banach spaces. Lower bounds for these coefficients, related to geometric constants introduced in Chapters IV and V, are obtained. The computation in direct sum spaces and L^p-spaces occupies the second part of the chapter.

  In Chapter VII we study fixed point theorems for Banach spaces which do not necessarily have normal structure, including nearly uniformly smooth Banach spaces. Basic tools for proving these theorems are Goebel-Karlovitz's lemma and Lin's lemma.

  In the setting of metric fixed point theory there is a kind of mappings which is of great interest: Uniformly Lipschitzian mappings, that is, mappings whose iterates are Lipschitzian with the same Lipschitz constant. A natural question arises: How small must the uniform Lipschitz constant be to assure the existence of a fixed point? In Chapter VIII we study this problem in connection with some of the geometric properties considered in previous chapters. A new geometric constant, the Lifshitz characteristic [L, 1975] is introduced, and its connection with certain other geometric constants is discussed. A fixed point theorem for uniformly Lipschitzian mappings related to Bynum's normal structure coefficient completes this study.

  In Chapter IX fixed point theorems for asymptotically regular mappings are discussed. To this end another geometric constant is introduced which plays for asymptotically regular mappings the role of the Lifshitz characteristic for uniformly Lipschitzian mappings. The study is carried out in a similar way to that in Chapter VIII, but in certain statements the Clarkson modulus of convexity is replaced by a modulus of near uniform convexity. In some sense, we could say that the results in this chapter are a noncompact version of those in Chapter VIII.

  The final Chapter is devoted to studying the rate of variation of the constants of ϕ-contractiveness of an operator when the measure of noncompactness is changed. Although some results in this direction were obtained by Nussbaum [N, 1970] and Webb [W1, 1973] for linear operators, the most significant results in this theory appeared after 1986 (see [AD], [ADL], [AD1], [AD2], [DL1], [Do1], [Do2], [DR1], [DR2], [Ro], [WZ], [Zh1]). The main tool in this study is the notion of the packing rate coefficient of a metric space. We compute the packing rate coefficients for ℓ^p and L^p-spaces, obtaining certain connections between the ϕ-contractive constants for different measures of noncompactness.

  Some results contained in this book have not been previously published. However, most results included are already known. As in any work of this type we have chosen to revise some proofs and include others heretofore unpublished. Nevertheless an effort has been made to indicate the original location of the theorems when known. Furthermore, biographic details for some of the late mathematicians quoted throughout the text (taken from [Ao], [Ed], [KN], [W], [De], [Ku2], [LO], [Kz] and [Sh]) have been included.

Chapter I

The Fixed Point Theorems of Brouwer and Schauder

We are going to dedicate the first chapter to the study of the fixed point theorem of Schauder [S, 1930]. We have divided the chapter into two parts: In the first part we give the finite dimensional version of Schauder's fixed point theorem (usually known as Brouwer's theorem [Br, 1912], though an equivalent form had been proved by Poincaré [Po, 1886]).

Luitzen Egbertus Jan Brouwer (1881–1967) was born on 27 February in Overschie (Holland), near Rotterdam. His father, Egbert, was village schoolmaster and lived to be 90. Brouwer attended primary school at Medemblik and Hoorn and a secondary school at Haarlem. He studied mathematics and science at the University of Amsterdam. From 1909 to 1912 he was a privaat-docent. In 1912 he was elected to the Chair for set theory, function theory and axiomatic at the University of Amsterdam, which he held till 1951. Also in 1912 he was elected Member of the Dutch Royal Academy of Science. In 1948 he was elected a Foreign Member of the Royal Society of London. He died on 2 December 1967 in a car accident at Blaricum near the house where he had lived for many years.

Brouwer worked intensely on problems of topology and analysis. In 1912, Brouwer demonstrated his famous fixed point theorem for continuous mappings, using some concepts of algebraic topology (a branch of mathematics created by Poincaré in 1895–1900). Another of the celebrated theorems proved by Brouwer was the theorem of the invariance of domain (1911–12): "If Ω is an open set of $\mathbb{R}^n$ and f is an injective continuous mapping from Ω into $\mathbb{R}^n$, then $f(\Omega)$ is also an open set".

However, Brouwer is better known as being the founder of modern intuitionism. The intuitionism philosophy found its origins at the end of the 19th century, when the rigour of both the numerical system and of geometry were subjects of major importance. Among its predecessors can be cited Kronecker and Poincaré.

From his Ph.D. (1907) "On the Foundations of Mathematics", Brouwer started to build his intuitionist philosophy and from 1918 he developed and generalized his points of view in a series of articles published in some of the most prestigious journals of the time. Among the most outstanding members of this school was Hermann Weyl.

In Brouwer's opinion, mathematical ideas were in the human mind before language, logic and experience. It was intuition which determined the validity and acceptability of ideas. These ideas violently clashed with those of the formalist school headed by Hilbert. For the formalists, the axiomatic method was more convenient for any precise research in whatever mathematical field. All that which could be the object of mathematical thought, stated Hilbert, was encompassed within the domain of the axiomatic method. In 1925 Brouwer vigorously attacked the formalists. Naturally, he said, the conventional axiomatic treatment will avoid contradictions, but in this way nothing of any mathematical value will be found. In turn Hilbert accused Brouwer of trying to throw overboard all that which was not in his own interest and called intuitionism the treason of science.

Many different proofs of this theorem can be found in the literature. This time we have chosen a proof using differential forms and Stoke's theorem (see [I2]). A different proof, strongly based on concepts of algebraic topology, can be found in [DG].

The second part of this chapter generalizes Brouwer's theorem to infinite dimensional Banach spaces via an approximation process. We give two alternative versions of this theorem following [Z]. In both versions, Schauder's theorem needs a hypothesis about compactness either on the operator or on its domain of definition.

1. The fixed point theorem of Brouwer and applications

Throughout this text $B(x, r)$ and $\overline{B}(x, r)$ denote respectively the open and closed ball centred at x with radius r in a normed linear space X, that is, $B(x, r) = \{y \in X : \|y - x\| < r\}$ and $\overline{B}(x, r) = \{y \in X : \|y - x\| \leq r\}$, where $\|.\|$ is the norm in X. Their boundary will be denoted by $S(x, r) = \{y \in X : \|y - x\| = r\}$. Whenever $X = \mathbb{R}^n$ with the euclidean norm, we will write $B_n(x, r)$, $\overline{B}_n(x, r)$ and $S_n(x, r)$ respectively. In this section $\|.\|$ will be the euclidean norm in $\mathbb{R}^n$ and $x \cdot y$ the inner product of the two vectors x and y.

We are going to prove one of the most important theorems in fixed point theory, the famous Brouwer's theorem, which shows that the euclidean unit ball $\overline{B}_n(0, 1)$ in $\mathbb{R}^n$ has the fixed point property for continuous functions, that is, if $f : \overline{B}_n(0, 1) \to \overline{B}_n(0, 1)$ is a continuous mapping, then there exists a point $x_0 \in \overline{B}_n(0, 1)$ such that $f(x_0) = x_0$.

DEFINITION 1.1. *Let X and X_1 be two topological spaces. X_1 is called a retract of X if*

(a) *X_1 is a subset of X.*

(b) *There is a continuous map $r : X \to X_1$ such that $r(x) = x$ for all x in X_1. The function r is called a retraction of X into X_1.*

The relationship between Brouwer's fixed point theorem and retractions is given in the following theorem:

THEOREM 1.2. *The Brouwer fixed point theorem is equivalent to the following assertion: "No indefinitely differentiable retraction exists from the euclidean unit ball $\overline{B}_n(0,1)$ onto the sphere $S_n(0,1)$".*

Proof. Let us suppose that Brouwer's theorem is true, but there is an indefinitely differentiable retraction r of $\overline{B}_n(0,1)$ onto $S_n(0,1)$. Then we define $r_1(x) = -r(x)$. Obviously r_1 is a continuous map of $\overline{B}_n(0,1)$ onto $S_n(0,1)$ which does not have a fixed point, contradicting Brouwer's theorem.

Conversely, let us suppose that there is no indefinitely differentiable retraction for the above pair of spaces. We shall first show that if $f : \overline{B}_n(0,1) \to \overline{B}_n(0,1)$ is any indefinitely differentiable map, then it has a fixed point. Indeed, if this were not the case we would have for each $x \in \overline{B}_n(0,1)$ that $x \neq f(x)$ and so we can consider the semiline beginning in $f(x)$ and passing through x. This semiline intersects $S_n(0,1)$ at a point, say $g(x)$. So, we have defined a map $g : \overline{B}_n(0,1) \to S_n(0,1)$ such that $g(x) \neq f(x)$ for all $x \in \overline{B}_n(0,1)$ and $g(x) = x$ for all $x \in S_n(0,1)$. Let us prove now that g is indefinitely differentiable. Indeed, for all $x \in \overline{B}_n(0,1)$ we have

$$g(x) = \alpha(x)x + (1 - \alpha(x))f(x)$$

where $\alpha(x)$ must be chosen such that the inner product $g(x) \cdot g(x) = 1$. This leads to the equation

$$\alpha(x)^2 \|x\|^2 + 2\alpha(x)(1 - \alpha(x))x \cdot f(x) + (1 - \alpha(x))^2 \|f(x)\|^2 = 1$$

which gives us $\alpha(x)$ as the solution of a second degree equation with indefinitely differentiable coefficients. It follows, by a well-known formula, that the solution has the same property. Thus the function $g(x)$ is an indefinitely differentiable retraction of $\overline{B}_n(0,1)$ onto $S_n(0,1)$.

Now, let $f : \overline{B}_n(0,1) \to \overline{B}_n(0,1)$ be a continuous map. The Weierstrass approximation theorem gives a sequence $\{f_n\}$ of indefinitely differentiable functions such that $\{f_n\}$ converges to f uniformly in $\overline{B}_n(0,1)$. Since $f_n : \overline{B}_n(0,1) \to \overline{B}_n(0,1)$ are indefinitely differentiable functions for all $n \in \mathbb{N}$, we obtain a sequence $\{x_n\}$ in $\overline{B}_n(0,1)$ such that $f_n(x_n) = x_n$. As $\{x_n\}$ lies inside the compact $\overline{B}_n(0,1)$ there is a subsequence (also denoted by $\{x_n\}$) convergent to a point $x_0 \in \overline{B}_n(0,1)$. It is now easy to conclude that $\lim_{n\to\infty} f_n(x_n) = f(x_0)$, and so x_0 is a fixed point for f. Thus the theorem is proved. $\square$

THEOREM 1.3 (BROUWER). *Let $f : \overline{B}_n(0,1) \to \overline{B}_n(0,1)$ be a continuous map. Then f has a fixed point.*

Proof. According to the previous theorem it suffices to show that no indefinitely differentiable retraction of $\overline{B}_n(0,1)$ onto $S_n(0,1)$ exists.

Suppose by use of a contradiction that such a retraction, say f, exists. We consider the differential form $\alpha = x_1 dx_2 \wedge \cdots \wedge dx_n$ and follow the notation of [Sp]. Then Stokes's theorem gives

$$\int_{S_n(0,1)} \alpha = \int_{f(S_n(0,1))} \alpha = \int_{S_n(0,1)} f^*\alpha = \int_{\overline{B}_n(0,1)} df^*\alpha$$

$$= \int_{\overline{B}_n(0,1)} f^* d\alpha = \int_{f(\overline{B}_n(0,1))} d\alpha = \int_{S_n(0,1)} d\alpha = 0.$$

But also Stokes's theorem gives

$$\int_{S_n(0,1)} \alpha = \int_{\overline{B}_n(0,1)} d\alpha = \text{volume}(\overline{B}_n(0,1)) > 0.$$

This contradiction proves the result. □

Remark 1.4. It is important to note that the proof of Brouwer's theorem is not constructive and does not give information about how to find the fixed points of the function.

Remark 1.5. In the case of only one variable, the Brouwer fixed point theorem is the following: "Every continuous function of $[-1,1]$ onto itself has a fixed point", or equivalently, "Every continuous function of $[-1,1]$ onto itself cuts the first quadrant diagonal at some point". In this case, the result is a very easy consequence of Bolzano's theorem. But even in $\mathbb{R}^2$ almost the full power of the above arguments are needed to verify the theorem.

In general, we cannot expect uniqueness of the fixed point in the theorem of Brouwer. So, we must consider the nonempty set $\mathcal{F}(f)$ of fixed points of f. By continuity this set is closed. It is a natural question to wonder what other properties it has. The following theorem [R] shows that no other special features can be inferred.

THEOREM 1.6. *Let F be a nonempty closed set contained in $\overline{B}_n(0,1)$. Then there exists a continuous function $f : \overline{B}_n(0,1) \to \overline{B}_n(0,1)$ such that $\mathcal{F}(f) = F$.*

Proof. For every $x \in \overline{B}_n(0,1)$ let $\mathrm{d}(x,F) = \inf\{\|x-y\| : y \in F\}$. Obviously this function is continuous. We now define $f : \overline{B}_n(0,1) \to \overline{B}_n(0,1)$ as

$$f(x) = \begin{cases} x - \mathrm{d}(x,F)\frac{x-x_o}{\|x-x_o\|} & \text{if } x \neq x_o \\ x_0 & \text{if } x = x_o \end{cases}$$

where x_0 is an arbitrary point of F. It is easy to show that this function is well defined and is continuous. Moreover $\mathcal{F}(f) = F$ and the theorem is proved. □

The Brouwer fixed point theorem is often used in the following more general version:

THEOREM 1.7. *If C is a nonempty, compact and convex subset of $\mathbb{R}^n$, and f is any continuous function of C into itself, then there is a fixed point for f.*

Proof. By multiplication and translation if necessary, we can suppose that C is contained in $\overline{B}_n(0,1)$. By the Projection Theorem (see [B, page 69, Proposition 1]), for all $x \in \overline{B}_n(0,1)$ there exists a unique point $P_C(x) \in C$ such that $\|x - P_C(x)\| = \inf_{y \in C} \|x - y\|$. It is well known that the map $P_C : \overline{B}_n(0,1) \to C$ is continuous and obviously satisfies $P_C(x) = x$ for every $x \in C$. Thus P_C is a retraction from $\overline{B}_n(0,1)$ onto C. Let us now consider the mapping $f \circ P_C : \overline{B}_n(0,1) \to C$. From Theorem 1.3 we have $x_0 \in \overline{B}_n(0,1)$ such that $f \circ P_C(x_0) = x_0$. Hence $x_0 \in C$ and $f(x_0) = x_0$. $\qquad\square$

COROLLARY 1.8. *Let $f : \mathbb{R}^n \to \mathbb{R}^n$ be a continuous mapping and suppose that, for some $r > 0$ and all $\lambda > 0$ we have $f(u) + \lambda u \neq 0$ for all u with $\|u\| = r$. Then there exists a point u_0, $\|u_0\| < r$ such that $f(u_0) = 0$.*

Proof. Suppose $f(u_0) \neq 0$ for every point $u_0 \in \overline{B}_n(0,r)$. Then the map $g : \overline{B}_n(0,r) \to \overline{B}_n(0,r)$ given by

$$g(u) = \frac{-f(u)r}{\|f(u)\|}$$

is continuous and well defined. So, Brouwer's theorem implies that there is a point $\overline{u} \in \overline{B}_n(0,r)$ such that $g(\overline{u}) = \overline{u}$ and this gives us that $f(\overline{u})r + \|f(\overline{u})\|\overline{u} = 0$. This equality contradicts the assumption on f since $\|\overline{u}\| = \|g(\overline{u})\| = r$. $\qquad\square$

Remark 1.9. This result, actually equivalent to the Brouwer fixed point theorem (see [I2, page 116, Theorem 4.2.2]), was proved first by Poincaré [Po, 1886], and some years later by Bohl [Bo, 1904]. It permits to prove the existence of a solution for the equation $f(u) = 0$ in $\overline{B}_n(0,r)$.

COROLLARY 1.10. *Let $f : \overline{B}_n(0,1) \to \mathbb{R}^n$ be a continuous mapping having the property that the euclidean norm of $f(x)$ is less or equal to one for all x with norm one. Then there exists $x_0 \in \overline{B}_n(0,1)$ such that $f(x_0) = x_0$.*

Proof. First, we are going to prove the result for f being indefinitely differentiable. Suppose by contradiction that $x \neq f(x)$ for all $x \in \overline{B}_n(0,1)$. We define the function

$$x \to g(x)$$

where $g(x)$ represents the point of norm one on the line that starts in $f(x)$ and passes through x. It is not difficult to show that g is an indefinitely differentiable retraction from $\overline{B}_n(0,1)$ onto $S_n(0,1)$, contradicting Brouwer's theorem.

If f is a continuous mapping, the Weierstrass approximation theorem permits us to find a sequence $\{f_n\}$ of indefinitely differentiable functions such that $\{f_n\}$ is uniformly convergent to f. Since the maps f_n are indefinitely differentiable for all $n \in \mathbb{N}$, we obtain a sequence $\{x_n\}$ in $\overline{B}_n(0,1)$ such that $f_n(x_n) = x_n$. As $\{x_n\}$ lies inside the compact set $\overline{B}_n(0,1)$ there is a subsequence (also denoted by $\{x_n\}$) convergent to a point $x_0 \in \overline{B}_n(0,1)$. It is now easy to conclude that $\lim_{n\to\infty} f_n(x_n) = f(x_0)$, and so x_0 is a fixed point for f. Thus the theorem is proved.   □

As an application of the Brouwer fixed point theorem we prove the following theorem concerning the existence of *periodic solutions to a differential equation*.

THEOREM 1.11. *Let $f : \overline{B}_n(0,1) \times \mathbb{R} \to \mathbb{R}^n$ be a mapping with continuous derivative at any point $(x,t) \in \overline{B}_n(0,1) \times \mathbb{R}$ and T- periodic in the second variable. Consider the differential equation*

$$x'(t) = f(x(t), t) \tag{1}$$

where x takes values in $\mathbb{R}^n$. If f satisfies the boundary condition $f(x,t) \cdot x < 0$ for every $x \in S_n(0,1)$ and any $t \in \mathbb{R}$, then the equation (1) has a periodic solution of period T.

Proof. The regularity and the boundary condition on f imply that for every initial value $x(t_0) = x_0 \in \overline{B}_n(0,1)$ the equation (1) has a unique solution in some neighbourhood of t_0 with values in $\overline{B}_n(0,1)$ (see [M, page 216, Theorem 3.1]) and with $x(t)$ depending continuously on $x(t_0)$. Using Zorn's lemma we can construct, by a standard argument, a maximal solution $x(t)$ on an interval J. If $t_1 \in \partial J$ the boundedness of f assures the existence of $\lim_{t\to t_1} x(t) = x_1$. Since the solution can be extended in a neighbourhood of (t_1, x_1) we deduce that $\partial J = \emptyset$, that is, $J = \mathbb{R}$.

So, for each $x_0 \in \overline{B}_n(0,1)$ there is a unique solution $x : \mathbb{R} \to \mathbb{R}^n$ of the differential equation with $x(0) = x_0$. Let $x_T = x(T)$. Then $x_T \in \overline{B}_n(0,1)$ and we can define a function $g : \overline{B}_n(0,1) \to \overline{B}_n(0,1)$ by $g(x_0) = x_T$. The map g is continuous, since $x(t)$ depends continuously on $x(0)$. Therefore, the Brouwer fixed point theorem implies that there exists $z \in \overline{B}_n(0,1)$ with $g(z) = z$. That means that the solution of the differential equation $x : \mathbb{R} \to \mathbb{R}^n$ given by $x(0) = z$ also satisfies $x(T) = z$. Now, we show that x has period T.

Consider the map $\Phi : \mathbb{R} \to \mathbb{R}^n$ defined by $\Phi(t) = x(t+T)$. It suffices to prove that $\Phi \equiv x$. In order to see this, we note that Φ is also a solution of the differential equation (1) with $\Phi(0) = z$. Indeed, for all $t \in \mathbb{R}$

$$\Phi'(t) = x'(t+T) = f(x(t+T), t+T) = f(\Phi(t), t).$$

Moreover,

$$\Phi(0) = x(0+T) = x(T) = z.   \qquad □$$

2. The fixed point theorem of Schauder and applications

Our goal is now to generalize the Brouwer fixed point theorem to infinite dimensional Banach spaces via an approximation process. We obtain the fixed point theorem of Schauder [S, 1930].

THEOREM 2.1 (SCHAUDER). *Let M be a nonempty, compact, convex subset of a Banach space X, and suppose $T : M \to M$ is a continuous mapping. Then T has a fixed point.*

Proof. Since $T(M)$ is compact, for each $n \in \mathbb{N}$ there exist elements $y_i \in T(M)$, $i = 1, \ldots, N$ such that

$$\min_i \|Tx - y_i\| < \frac{1}{n}$$

for all $x \in M$. We now consider the so-called *Schauder operator*, defined for all $x \in M$ by

$$P_n(x) = \frac{\sum_{i=1}^{N} a_i(x) y_i}{\sum_{i=1}^{N} a_i(x)}$$

where $a_i(x) = \max\{1/n - \|Tx - y_i\|, 0\}$. This operator has the following properties:

(i) P_n is a continuous map, because the continuous functions a_i do not all vanish simultaneously for $x \in M$.

(ii) $\sup_{x \in M} \|Tx - P_n(x)\| \leq 1/n$. Indeed, as $a_i(x) = 0$ unless $\|y_i - Tx\| < 1/n$ we have that

$$\|P_n(x) - Tx\| = \left\| \frac{\sum_i a_i(x)(y_i - Tx)}{\sum_i a_i(x)} \right\| \leq \frac{\sum_i a_i(x)\frac{1}{n}}{\sum_i a_i(x)} = \frac{1}{n}$$

for all $x \in M$.

(iii) The dimension of $\operatorname{span}(P_n(M))$ is finite because $\{y_1, y_2, \ldots, y_N\}$ is a generating system for $P_n(M)$.

Let $M_n = \overline{co}(\{y_1, \ldots, y_N\})$, where for any set A, $\overline{co}(A)$ denotes the closed convex hull of A. The convexity of M implies that $M_n \subset \overline{co}(T(M)) \subset M$. Moreover M_n is a convex, compact and finite dimensional set and $P_n(x) \in M_n$ for all $x \in M_n$. So, by the Brouwer fixed point theorem, there exists a fixed point $x_n = P_n(x_n)$, where $x_n \in M_n \subset M$.

Since M is compact, there is a convergent subsequence, again denoted by $\{x_n\}$, such that $\{x_n\}$ converges to $x \in M$. This x is the desired fixed point, since

$$\|x_n - Tx\| \leq \|P_n(x_n) - Tx_n\| + \|Tx_n - Tx\|$$

and the right-hand side of the inequalities vanishes as $n \to \infty$ by (ii) and the continuity of T. So $x = Tx$ and the proof is concluded. $\square$

Remark 2.2. As is well known, when we pass to infinite dimensional spaces, there are bounded and closed subsets which are not compact. So, it is natural to wonder if Schauder's theorem holds if M is only a convex, closed and bounded subset of X. The following *example of Kakutani* [K, 1943] provides a strong negative answer to this question.

PROPOSITION 2.3. *There is a fixed point free continuous mapping on the unit ball of $\ell^2(\mathbb{Z})$.*

Proof. We consider $\ell^2(\mathbb{Z})$ with the standard basis consisting of the sequence $\{e_n : n \in \mathbb{Z}\}$ where $e_n = (\dots, 0, 0, 1, 0, 0, \dots)$ with the one in position n, and denote by B the closed unit ball in this space. For $x \in \ell^2(\mathbb{Z})$ we can write

$$x = (\dots, x^{-1}, x^0, x^1, x^2, \dots) = \sum_n x^n e_n.$$

We now define the right shift operator $U : \ell^2(\mathbb{Z}) \to \ell^2(\mathbb{Z})$ by $Ux = \sum_n x^n e_{n+1}$.

The relation

$$x - Ux = \sum_n (x^n - x^{n-1})e_n = ce_0$$

requires that $x^n = x^0$ for all $n > 0$ and that $x^n = x^{-1}$ for all $n < 0$. For a point of $\ell^2(\mathbb{Z})$ this is only possible if $x^0 = x^{-1} = 0$. So, $x - Ux$ is a multiple of e_0 if and only if $x = 0$.

Let us now consider the mapping defined by

$$Tx = (1 - \|x\|)e_0 + Ux.$$

T is continuous and maps B into B, since if $\|x\| \leq 1$ we have

$$\|Tx\| \leq (1 - \|x\|)\|e_0\| + \|Ux\| = (1 - \|x\|) + \|x\| = 1.$$

Finally, T is a fixed point free mapping. Indeed, if

$$x = Tx = (1 - \|x\|)e_0 + Ux$$

then $x - Ux = (1 - \|x\|)e_0$ which is clearly impossible if $x = 0$ and is impossible, as we have seen above, if $x \neq 0$. $\square$

Remark 2.4. Though T is fixed point free, it is not difficult to check that $\inf\{\|x - Tx\| : x \in B\} = 0$. Indeed, it suffices to consider the sequence $\{x_n\}$ of elements of $\ell^2(\mathbb{Z})$ given by

$$x_n = (\dots, 0, \overset{(-n}{\frac{1}{\sqrt{2n}}}, \dots, \overset{(n-1}{\frac{1}{\sqrt{2n}}}, 0, \dots)$$

such that $\|x_n\| = 1$ and $\|x_n - Tx_n\| = 1/\sqrt{n}$ for every $n \in \mathbb{N}$.

In the following example we show that the number $\inf\{\|x - Tx\| : x \in B\}$ can be greater than 0.

Example 1: Let c_0 be the Banach space of all real sequences convergent to zero with the supremum norm and B be the unit ball of this space. Fix $k > 1$ and consider the mapping $a : [-1, 1] \to \mathbb{R}$ given by $a(t) = \min\{1, k|t|\}$. Let us now define the function $T : B \to B$ as $Tx = (1, a(x^1), a(x^2), \dots)$ for every $x = (x^1, x^2, \dots) \in B$.

Obviously T is a well defined and continuous mapping. Moreover $\|x - Tx\| > 1 - 1/k$ for all $x \in B$. Indeed, if this were not the case for some $x \in B$ we would have $\|x - Tx\| \leq 1 - 1/k$ which implies $x^n \geq 1/k$ for every $n \in \mathbb{N}$, contradicting $x \in c_0$.

It is possible to prove a stronger result than the one in Proposition 2.3. In [Kl] it was shown that every convex noncompact set in a Banach space lacks the fixed point property for continuous functions. In [LS] the same result is proved for Lipschitzian mappings. Therefore, if we want to obtain a result like Theorem 2.1 without requiring compactness of the set M, we must compensate for the loss of compactness of the set by a stronger condition than continuity for the operator T. Before giving this new version of Schauder's theorem, we need to introduce the concept of a compact operator.

DEFINITION 2.5. *Let X and Y be Banach spaces and M a subset of X. A mapping $T : M \subset X \to Y$ is called compact if T is continuous and maps bounded sets into relatively compact sets.*

Remark 2.6. Compact operators are very useful in nonlinear functional analysis. Many results about continuous operators on $\mathbb{R}^n$ are generalized to Banach spaces by replacing "continuous" with "compact". Let us see two examples of compact operators.

Example 2: For finite dimensional Banach spaces, continuous and compact operators are the same whenever the domain of definition is closed. Indeed, if M is a bounded set, then $\overline{M}$ is compact. Thus $f(\overline{M})$ is also compact, and so $f(M)$ is relatively compact.

Example 3: Integral operators with sufficiently regular kernels provide the most important examples of nonlinear compact operators on infinite dimensional Banach spaces. A wide study of these operators may be found in [M, Chapter 5]. We are showing here two classic examples.

Suppose we have a continuous function

$$K : [a, b] \times [a, b] \times [-R, R] \to \mathbb{R}$$

where $-\infty < a < b < +\infty$, $0 < R < +\infty$, and set

$$M = \{x \in C([a, b]) : \|x\| \leq R\},$$

where $C([a, b])$ is the Banach space of all real continuous functions defined on the interval $[a, b]$ with the norm $\|x\| = \max\{|x(t)| : t \in [a, b]\}$. Then, the integral operators S and T defined by

$$(Tx)(t) = \int_a^b K(t, s, x(s))ds \quad t \in [a, b]$$

$$(Sx)(t) = \int_a^t K(t, s, x(s))ds \quad t \in [a, b]$$

map M into $C([a,b])$ and are compact. These integral operators are called, respectively, *Fredholm and Volterra operators*.

Proof. We prove the result for T. The other case is similar.

(i) Since the set $A = [a,b] \times [a,b] \times [-R,R]$ is compact and the mapping K is continuous, we obtain that K is bounded and uniformly continuous on A. Therefore there is a number α such that $|K(t,s,x)| \leq \alpha$, for all $(t,s,x) \in A$, and for every $\varepsilon > 0$ there is a $\delta(\varepsilon) > 0$ such that

$$|K(t_1,s_1,x_1) - K(t_2,s_2,x_2)| < \varepsilon$$

for all $(t_i,s_i,x_i) \in A$, $i = 1,2$, satisfying $|t_1 - t_2| + |s_1 - s_2| + |x_1 - x_2| < \delta(\varepsilon)$.

(ii) Let $x \in M$ and $z = Tx$. We show that $z \in C([a,b])$. Indeed, for $|t_1 - t_2| < \delta(\varepsilon)$, we have the inequality

$$|z(t_1) - z(t_2)| = \left| \int_a^b K(t_1,s,x(s))ds - \int_a^b K(t_2,s,x(s))ds \right|$$

$$\leq \int_a^b |(K(t_1,s,x(s)) - K(t_2,s,x(s)))|ds \leq (b-a)\varepsilon.$$

So, $z = Tx$ is uniformly continuous on $[a,b]$. Thus T maps M into $C([a,b])$. Moreover

$$|z(t)| \leq \left| \int_a^b K(t,s,x(s))ds \right| \leq (b-a)\alpha$$

for all $t \in [a,b]$.

(iii) The inequalities in (ii) hold uniformly for all $z = Tx$ whenever $x \in M$. It follows from the Arzela-Ascoli theorem that the set $T(M)$ is relatively compact.

(iv) The operator T is continuous on M. Indeed, given $\varepsilon > 0$ we take $\delta(\varepsilon)$ as above. Then, if x and y are in M with $\|x - y\| < \delta(\varepsilon)$ we have

$$\|Tx - Ty\| = \max_{a \leq t \leq b} |Tx(t) - Ty(t)|$$

$$= \max_{a \leq t \leq b} \left| \int_a^b (K(t,s,x(s)) - K(t,s,y(s)))ds \right| < (b-a)\varepsilon.$$

From (iii) and (iv) we imply the compactness of T. $\qquad\square$

We can now give a new version of the Schauder fixed point theorem. This version is more frequently used in applications and M is often chosen to be a ball.

THEOREM 2.7 (ALTERNATE VERSION OF THE SCHAUDER FIXED POINT THEOREM). *Let M be a nonempty, convex, closed, bounded subset of a Banach space X, and suppose $T : M \rightarrow M$ is a compact operator. Then T has a fixed point.*

Proof. Let $A = \overline{\mathrm{co}}(T(M))$. Then $A \subset M$ and the set A is obviously convex. Furthermore, A is compact by Mazur's theorem (see [DS, Vol.I, page 406, Theorem V.2.6]) and $T(A) \subset A$. Thus the restriction $T : A \to A$ has a fixed point by Theorem 2.1. This point is also a fixed point of T in M. □

The Schauder fixed point theorem is a very useful tool for proving the existence of solutions to many nonlinear problems, especially problems concerning ordinary and partial differential equations. We illustrate this by using the Schauder theorem to prove the following *generalized Peano's theorem* [Co, 1957].

THEOREM 2.8. *Let a and b be positive real numbers, J the real interval $[t_0 - a, t_0 + a]$, $D = \overline{B}(x_0, b)$ in a Banach space X and $f : J \times D \to X$ a mapping. Let us consider the problem of Cauchy*

$$x'(t) = f(t, x(t)), \quad x(t_0) = x_0.$$

Then, if f is a compact operator, there exists at least one solution to this problem. This solution is at least defined in the interval $[t_0 - h, t_0 + h]$, where $h = \min\{a, b/M\}$ and $M = \sup\{\|f(x, y)\| : (x, y) \in J \times D\}$.

Proof. Instead of the proposed initial value problem we can consider the equivalent integral equation

$$x(t) = x_0 + \int_{t_0}^{t} f(s, x(s))ds.$$

Let $M = \sup\{\|f(t, x)\| : (t, x) \in J \times D\}$. Since $f(J \times D)$ is relatively compact, the number $M < +\infty$. Let $h = \min\{a, b/M\}$ and $I = [t_0 - h, t_0 + h]$.

Let us consider the Banach space $\mathcal{C}(I, X)$ of all continuous functions x from I into X with the supremum norm and define the operator

$$T : \mathcal{C}(I, D) \to \mathcal{C}(I, D)$$

as

$$Tx(t) = x_0 + \int_{t_0}^{t} f(s, x(s))ds$$

for all $x \in \mathcal{C}(I, D)$ and $t \in I$. It is not difficult to show that T is well defined and that $\mathcal{C}(I, D)$ is a closed, bounded and convex subset of $\mathcal{C}(I, X)$.

We show that T is a compact operator:

(I) T is continuous on $\mathcal{C}(I, D)$.

Indeed, let $\{x_n\}$ be a sequence in $\mathcal{C}(I, D)$ such that $\|x_n - x\| \to 0$. Then

$$\|Tx_n - Tx\| = \sup_{t \in I} \|Tx_n(t) - Tx_(t)\| \leq h \sup_{s \in I} \|f(s, x_n(s)) - f(s, x(s))\| \to 0$$

as $n \to \infty$. Indeed, otherwise there would be an $\varepsilon_0 > 0$ and a sequence $\{s_n\}$ in I for which

$$\|f(s_n, x_n(s_n)) - f(s_n, x(s_n))\| \geq \varepsilon_0. \tag{2}$$

As $\{s_n\}$ is in the compact I, there is a subsequence (also denoted by $\{s_n\}$) convergent to a point $s_0 \in I$. Moreover

$$\|x_n(s_n) - x(s_0)\| \leq \|x_n(s_n) - x(s_n)\| + \|x(s_n) - x(s_0)\| \to 0$$

as $n \to \infty$. Therefore, both $f(s_n, x_n(s_n))$ and $f(s_n, x(s_n))$ converge to $f(s_0, x(s_0))$ as $n \to \infty$, contradicting (2).

(II) $T(\mathcal{C}(I, D))$ is relatively compact.

We will utilize the Arzela-Ascoli theorem. So we must prove:

(II.a) $T(\mathcal{C}(I, D))$ is an equicontinuous set.

Indeed, let $t_0 \in I$ and $\varepsilon > 0$. If we choose $\delta = \varepsilon/2M$ then
$\|Tx(t) - Tx(t_0)\| = \|\int_{t_0}^t f(s, x(s))ds\| \leq M|t_0 - t| < \varepsilon$ for all $x \in \mathcal{C}(I, D)$ and $t \in I$ with $|t - t_0| < \delta$.

(II.b) The set $\{Tx(t) : x \in \mathcal{C}(I, D)\}$ is relatively compact in X for every $t \in I$.

We shall utilize the following generalized theorem of mean value [M, page 26, Theorem 2.1]:

"Let J be a real interval, X a Banach space and $f : J \to X$ a differentiable map. Let $\alpha, \beta \in J$ with $\alpha < \beta$. Then $f(\beta) - f(\alpha) \in (\beta - \alpha)\overline{\text{co}}(\{f'(t) : t \in [\alpha, \beta]\})$."

In our case $J = I$ and $f(t) = Tx(t)$. It is clear that the set $K = \{f(t, x(t)) : t \in I, x \in \mathcal{C}(I, D)\}$ is relatively compact and so $\overline{\text{co}}(K)$ is a compact set. Moreover for $\alpha = t_0$ and $\beta = t$ we have

$$Tx(t) - Tx(t_0) = \int_{t_0}^t f(s, x(s))ds \in (t - t_0)\overline{\text{co}}(\{f(s, x(s)) : s \in [t_0, t]\})$$
$$\subset (t - t_0)\overline{\text{co}}(K)$$

and therefore for every $t \in I$ and $x \in \mathcal{C}(I, D)$ we have that $Tx(t)$ is contained in the compact set $x_0 + (t - t_0)\overline{\text{co}}(K)$. It follows that the set $\{Tx(t) : x \in \mathcal{C}(I, D)\}$ is relatively compact for all $t \in I$.

Now, the Schauder fixed point theorem implies the existence of a solution $x = Tx$ with $x \in \mathcal{C}(I, D)$. This function x is a solution of the integral equation and hence of the initial value problem. $\qquad\square$

Remark 2.9. If $X = \mathbb{R}^n$ the condition "f is a compact map" is automatically verified since f is continuous on the compact set $J \times D$. So, Theorem 2.8 reduces to the well known theorem of Peano [P, 1890]. If X is an infinite dimensional Banach space, then Peano's theorem cannot be applied. A fundamental argument in the proof of Peanos' theorem requires that bounded sets be relatively compact, and as we know this is no longer in infinite dimensional Banach spaces. In fact, Godunov showed [G, 1975] that in every infinite dimensional Banach space there is a continuous function such that the corresponding initial value problem has no solution.

Chapter II

Measures of Noncompactness

As we have seen in Chapter I, compactness plays an essential role in the proof of the Schauder fixed point theorem. However, there are some important problems where the operators are not compact.

The first step to extend the Schauder theorem to noncompact operators was given by G. Darbo [D, 1955]. The main idea is to define a new class of operators which map any bounded set in a "more compact" set. In order to state the property: A set is mapped into a "more compact" set, we need to define some "measure of noncompactness". The first such measure was defined by Kuratowski [Ku1, 1930] in connection with certain problems of General Topology. If B is a bounded set of a metric space, the measure of noncompactness of B is defined by

$$\alpha(B) = \inf \left\{ \begin{array}{l} \varepsilon > 0 : B \text{ can be covered by} \\ \text{finitely many sets with diameter} \leq \varepsilon \end{array} \right\} .$$

Kazimierz Kuratowski (1896–1980) was born in Warsaw (Poland) on 2 February. He took his Ph.D. in mathematics at Warsaw University in 1927, was appointed professor at the Lwów Technical University in 1927 and professor at the University of Warsaw in 1934. A great mathematician and one of the creators of modern topology, he had an enormous influence on research and education, not only in Poland but also throughout the whole world via his many students. After the end of World War II his great achievement was the reestabilishment and reorganization of numerous mathematical activities in Poland. He became a member of the Polish Academy of Learning in 1945 and of the Polish Academy of Science since its foundation in 1952. He was one of the founders of "Fundamenta Mathematica" and editor of this important journal until his death. Of his 172 publications the most important are devoted to set theory and topology, including the fundamental treatise on set-theoretic topology titled "Topology" ([Ku3 and Ku4], first ed. 1934) and the university text "An introduction to Set Theory and Topology" ([Ku5], first ed. 1954).

Darbo used this measure to generalize Schauder's theorem to a wide class of operators, called k-set-contractive operators, which satisfy the condition $\alpha(T(A)) \leq k\alpha(A)$ for some $k \in [0, 1)$. In 1967 Sadovskiĭ [Sa1] generalized Darbo's theorem to set-condensing operators.

Other measures of noncompactness have been defined since then. The most important ones are the measure of noncompactness of Hausdorff

$$\chi(B) = \inf \left\{ \begin{matrix} \varepsilon > 0 : B \text{ can be covered by} \\ \text{finitely many balls with radii } \leq \varepsilon \end{matrix} \right\}$$

introduced by Gohberg, Gol'denshteĭn and Markus [GGM, 1957] and the separation measure of noncompactness

$$\beta(B) = \sup\{r > 0 : B \text{ has an infinite } r\text{-separation}\}$$

considered by Istrăţescu [I1, 1972], Sadovskiĭ [Sa2, 1968] and other authors.

Measures of noncompactness are very useful tools in the theory of operator equations in Banach spaces. They are very often used in the theory of functional equations, including ordinary differential equations, equations with partial derivatives, integral and integro-differential equations, optimal control theory, etc. In particular, the fixed point theorems derived from them have many applications. There exists a considerable literature devoted to this subject (see for example [I2], [BG], [Z] and [AKPRS]).

In recent years measures of noncompactness have also been utilized to define new geometrical properties of Banach spaces which are interesting for fixed point theory (see for instance [Ro1], [Ro2], [Mo], [KMP], [KP], [Ku], [Ba], [ADF1], [ADF2], [Se1], [Pr5] and references therein). We shall study some of these properties in the next chapters.

1. The general notion of a measure of noncompactness

In this section we axiomatize the notion of a measure of noncompactness on a metric space. It seems that the axiomatic approach is the best way of dealing with measures of noncompactness. Obviously, it is possible to use several systems of axioms which are not necessarily equivalent. The set of axioms should satisfy two requirements: first, it should have natural realizations and second, it should provide useful tools for applications.

In the books of [BG] and [AKPRS] two different patterns for axiomatically introducing measures of noncompactness in Banach spaces are presented. However, the notion of a measure of noncompactness was originally introduced in metric spaces and we are going to give our axiomatic definition for this class of spaces.

DEFINITION 1.1. *Let (X, d) be a complete metric space and $\mathcal{B}$ the family of bounded subsets of X. A map*

$$\phi : \mathcal{B} \to [0, +\infty)$$

is called a measure of noncompactness (MNC) defined on X if it satisfies the following properties:

(a) *Regularity:* $\phi(B) = 0 \Leftrightarrow B$ *is a precompact set.*
(b) *Invariant under closure:* $\phi(B) = \phi(\bar{B}), \quad \forall B \in \mathcal{B}.$
(c) *Semi-additivity:* $\phi(B_1 \cup B_2) = \max\{\phi(B_1), \phi(B_2)\}, \quad \forall B_1 \in \mathcal{B}, \forall B_2 \in \mathcal{B}.$

From these axioms, we can immediately deduce the following properties:

(1) Monotonicity: $B_1 \subset B \Rightarrow \phi(B_1) \leq \phi(B).$
(2) $\phi(B_1 \cap B_2) \leq \min\{\phi(B_1), \phi(B_2)\}, \quad \forall B_1 \in \mathcal{B}, \forall B_2 \in \mathcal{B}.$
(3) Non-singularity: If B is a finite set, then $\phi(B) = 0.$
(4) *Generalized Cantor's intersection theorem:* If $\{B_n\}$ is a decreasing sequence of nonempty, closed and bounded subsets of X and $\lim_{n \to \infty} \phi(B_n) = 0$, then the intersection B_∞ of all B_n is nonempty and compact.

Proof. We only prove (4). Let $\{x_n\}$ be a sequence such that $x_n \in B_n$ for all $n \in \mathbb{N}$ and consider the decreasing sequence of sets $\{C_n\}$ given by $C_n = \{x_i : i \geq n\}$. Obviously $C_n \subset B_n$ and $\phi(C_1) = \phi(C_n) \leq \phi(B_n)$ for every $n \in \mathbb{N}$.

Since $\lim_{n \to \infty} \phi(B_n) = 0$, it follows that $\phi(C_1) = 0$ and so $\{x_n\}$ is a precompact set. Let $\bar{x}$ be the limit of a subsequence of $\{x_n\}$. Obviously $\bar{x} \in B_n$ for all $n \in \mathbb{N}$ and hence $B_\infty \neq \emptyset$.

Moreover, as $\phi(B_\infty) \leq \phi(B_n)$ for every $n \in \mathbb{N}$ and $\lim_{n \to \infty} \phi(B_n) = 0$, we obtain $\phi(B_\infty) = 0$ and so B_∞ is compact since it is a closed set. $\square$

Moreover, if X is a Banach space, the measure of noncompactness ϕ can enjoy some additional properties. Let us list some of them:

(5) Semi-homogeneity: $\phi(tB) = |t|\phi(B)$ for any number t and $B \in \mathcal{B}.$
(6) Algebraic semi-additivity: $\phi(B_1 + B_2) \leq \phi(B_1) + \phi(B_2), \quad \forall B_1 \in \mathcal{B}, \forall B_2 \in \mathcal{B}.$
(7) Invariance under translations: $\phi(x_0 + B) = \phi(B)$ for any $x_0 \in X$ and $B \in \mathcal{B}.$
(8) Lipschitzianity: $|\phi(B_1) - \phi(B_2)| \leq L_\phi \rho(B_1, B_2)$, where ρ denotes the *Hausdorff semimetric* $\rho(B_1, B_2) = \inf\{\varepsilon > 0 : B_2 \subset B_1 + \varepsilon\overline{B}(0,1), B_1 \subset B_2 + \varepsilon\overline{B}(0,1)\}.$
(9) Continuity: For every $B \in \mathcal{B}$ and for all $\varepsilon > 0$, there is $\delta > 0$ such that $|\phi(B) - \phi(B_1)| < \varepsilon$ for all B_1 satisfying $\rho(B, B_1) < \delta.$
(10) Invariance under passage to the convex hull: $\phi(\operatorname{co}(B)) = \phi(B)$ for all $B \in \mathcal{B}.$

Example 1: In every metric space X, the map

$$\phi(B) = \begin{cases} 0 & \text{if } B \text{ is precompact} \\ 1 & \text{otherwise} \end{cases}$$

is a measure of noncompactness, which will be called the *discrete measure of noncompactness*. This measure is algebraically semi-additive and invariant under translations and passage to the convex hull; it is neither semihomogeneous nor continuous. In the next section we are going to define two more interesting measures of noncompactness.

2. The Kuratowski and Hausdorff measures of noncompactness

In this section we define the Kuratowski and Hausdorff measures of noncompactness (MNCs) and study their basic properties. As we know, if $B \in \mathcal{B}$ is not a precompact set, there is a number $\varepsilon > 0$ such that B cannot be covered by finitely many sets (or balls) with diameter $\leq \varepsilon$. Hence, we can give the following definition:

DEFINITION 2.1. *Let (X, d) be a complete metric space and $\mathcal{B}$ the family of bounded subsets of X. For every $B \in \mathcal{B}$ we define the mappings α (of Kuratowski) and χ (of Hausdorff) in the following way:*

$$\alpha(B) = \inf\{\varepsilon > 0 : B \text{ can be covered by finitely many sets of diameter} \leq \varepsilon\}$$
$$\chi(B) = \inf\{\varepsilon > 0 : B \text{ can be covered by finitely many balls of radius} \leq \varepsilon\}$$

Remark 2.2.

(a) As usual, the *diameter of a set B* is the number $\sup\{d(x, y) : x \in B, y \in B\}$ denoted by $\operatorname{diam}(B)$, with $\operatorname{diam}(\emptyset) = 0$. It is clear that $0 \leq \alpha(B) \leq \operatorname{diam}(B) < +\infty$ for each nonempty bounded subset B of X and that $\operatorname{diam}(B) = 0$ if and only if B is an empty set or consists of exactly one point. Some other important properties of the diameter are the following:

(i) If $B_1 \subset B_2$ then $\operatorname{diam}(B_1) \leq \operatorname{diam}(B_2)$;

(ii) $\operatorname{diam}(\overline{B}) = \operatorname{diam}(B)$;

(iii) *Cantor's intersection theorem:* If $\{B_n\}$ is a decreasing sequence of nonempty, closed and bounded subsets of X and $\lim_{n \to \infty} \operatorname{diam}(B_n) = 0$, then the intersection B_∞ of all B_n is nonempty and consists of exactly one point.

Moreover, if X is a Banach space, then:

(iv) $\operatorname{diam}(tB) = |t| \operatorname{diam}(B)$ for any real number t;

(v) $\operatorname{diam}(x + B) = \operatorname{diam}(B)$ for any $x \in X$;

(vi) $\operatorname{diam}(B_1 + B_2) \leq \operatorname{diam}(B_1) + \operatorname{diam}(B_2)$;

(vii) $\operatorname{diam}(\operatorname{co}(B)) = \operatorname{diam}(B)$.

Indeed, parts (i)-(vi) are routine. To see (vii) let $x, y \in \operatorname{co}(B)$. Then $x = \sum_{i=1}^{n} t_i x_i$ and $y = \sum_{j=1}^{m} s_j y_j$ with $x_i, y_j \in B$, $\sum_{i=1}^{n} t_i = 1$ and $\sum_{j=1}^{m} s_j = 1$. Thus

$$\|x - y\| = \left\| \sum_{i=1}^{n} t_i x_i - \sum_{j=1}^{m} s_j y_j \right\| = \left\| \sum_{j=1}^{m} \sum_{i=1}^{n} s_j t_i x_i - \sum_{i=1}^{n} \sum_{j=1}^{m} t_i s_j y_j \right\| \leq$$

$$\leq \sum_{j=1}^{m} \sum_{i=1}^{n} t_i s_j \|x_i - y_j\| \leq \operatorname{diam}(B) \sum_{j=1}^{m} \sum_{i=1}^{n} t_i s_j = \operatorname{diam}(B)$$

and it follows that $\operatorname{diam}(\operatorname{co}(B)) \leq \operatorname{diam}(B)$. Since the opposite inequality is obvious, the assertion of (vii) follows.

(b) Recall that in a Banach space X, a set $S \subset X$ is called an ε-net of B if $B \subset S + \varepsilon \overline{B}(0,1) = \{s + \varepsilon b : s \in S, b \in \overline{B}(0,1)\}$. So, the definition of the χ-measure in Banach spaces is equivalent to the following:

$$\chi(B) = \inf\{\varepsilon > 0 : B \text{ has a finite } \varepsilon - \text{net}\}.$$

(c) Obviously, in both definitions we can replace the inequalities $\leq$ by $<$.

The next properties are common to α and χ and so we are going to use ϕ to denote either of them. These properties follow immediately from the definitions and show that both mappings are measures of noncompactness in the sense of Definition 1.1.

PROPOSITION 2.3. *Let ϕ denote α or χ. Then the following properties are satisfied in any complete metric space X:*

(a) *Regularity: $\phi(B) = 0$ if and only if B is precompact.*

(b) *Invariant under passage to the closure: $\phi(\bar{B}) = \phi(B)$ for all $B \in \mathcal{B}$.*

(c) *Semi-additivity: $\phi(B_1 \cup B_2) = \max\{\phi(B_1), \phi(B_2)\}$, $\quad \forall B_1 \in \mathcal{B}, \forall B_2 \in \mathcal{B}$.*

(d) *Monotonicity: $B_1 \subset B_2$ implies $\phi(B_1) \leq \phi(B_2)$.*

(e) *$\phi(B_1 \cap B_2) \leq \min\{\phi(B_1), \phi(B_2)\}$, $\quad \forall B_1 \in \mathcal{B}, \forall B_2 \in \mathcal{B}$.*

(f) *Non-singularity: If B is a finite set, then $\phi(B) = 0$.*

(g) *Generalized Cantor's intersection theorem: If $\{B_n\}$ is a decreasing sequence of nonempty, closed and bounded subsets of X and $\lim_{n \to \infty} \phi(B_n) = 0$, then the intersection B_∞ of all B_n is nonempty and compact.*

If X is a Banach space, then we also have:

(h) *Semi-homogeneity: $\phi(tB) = |t|\phi(B)$ for any real number t and $B \in \mathcal{B}$.*

(i) *Algebraic semi-additivity: $\phi(B_1 + B_2) \leq \phi(B_1) + \phi(B_2)$, $\quad \forall B_1 \in \mathcal{B}, \forall B_2 \in \mathcal{B}$.*

(j) *Permanence under translations: $\phi(x_0 + B) = \phi(B)$ for any $x_0 \in X$ and $B \in \mathcal{B}$.*

(k) *Lipschitzianity: $|\phi(B_1) - \phi(B_2)| \leq L_\phi \rho(B_1, B_2)$, where $L_\chi = 1$, $L_\alpha = 2$ and ρ denotes the Hausdorff semimetric.*

(l) *Continuity: For every $B \in \mathcal{B}$ and for all $\varepsilon > 0$, there is $\delta > 0$ such that $|\phi(B) - \phi(B_1)| < \varepsilon$ for all B_1 satisfying $\rho(B, B_1) < \delta$.*

Some less trivial properties of these measures of noncompactness are obtained in the next theorems.

THEOREM 2.4. *The Kuratowski and Hausdorff MNCs are invariant under passage to the convex hull: $\phi(B) = \phi(\text{co}(B))$.*

Proof. We are going to prove only that $\alpha(B) = \alpha(\text{co}(B))$. The proof for χ is analogous.

Since $B \subset \text{co}(B)$ we obtain $\alpha(B) \leq \alpha(\text{co}(B))$. Conversely, let us show that $\alpha(\text{co}(B)) \leq \alpha(B)$. Indeed, for every $\varepsilon > 0$ there exists a finite cover $\{B_1, B_2, \ldots, B_n\}$ of B such that $\text{diam}(B_i) \leq \alpha(B) + \varepsilon$ for all $i = 1, 2, \ldots, n$. We can assume that every B_i is a convex set since $\text{diam}(\text{co}(B)) = \text{diam}(B)$.

Let us define

$$\sigma = \left\{ (\lambda_1, \ldots, \lambda_n) \in \mathbb{R}^n : \sum_{i=1}^{n} \lambda_i = 1, \ \lambda_i \geq 0 \ \forall i = 1, \ldots, n \right\}$$

and $A(\lambda) = \sum_{i=1}^{n} \lambda_i B_i$ for every $\lambda \in \sigma$.

It follows from Proposition 2.3 that

$$\alpha(A(\lambda)) \leq \sum_{i=1}^{n} \lambda_i \alpha(B_i) \leq \alpha(B) + \varepsilon.$$

We show now that the set $\bigcup_{\lambda \in \sigma} A(\lambda)$ is convex.

Let $z = tx + (1-t)y$ and $\eta = t\lambda + (1-t)\mu$. It suffices to prove that if $0 \leq t \leq 1$, $x \in A(\lambda)$ and $y \in A(\mu)$, then $z \in A(\eta)$. Indeed, let $x = \sum_{i=1}^{n} \lambda_i x_i$ and $y = \sum_{i=1}^{n} \mu_i y_i$ where $\lambda = (\lambda_1, \ldots, \lambda_n) \in \sigma$, $\mu = (\mu_1, \ldots, \mu_n) \in \sigma$ and $x_i, y_i \in B_i$ for all $i = 1, \ldots, n$.

It is not difficult to check that the point z can be written in the form

$$z = \sum_{i=1}^{n} \eta_i z_i$$

where $z_i = \rho_i x_i + (1 - \rho_i) y_i$ and

$$\rho_i = \begin{cases} t\lambda_i/\eta_i & \text{for } \eta_i > 0 \\ 0 & \text{for } \eta_i = 0 \end{cases}$$

So, $z \in A(\eta)$ since $\eta \in \sigma$ and $z_i \in B_i$ because B_i is a convex set.

We are now ready to prove the result.

Since $B \subset \bigcup_{i=1}^{n} B_i \subset \bigcup_{\lambda \in \sigma} A(\lambda)$ and the set $\bigcup_{\lambda \in \sigma} A(\lambda)$ is convex, it follows that $\text{co}(B) \subset \bigcup_{\lambda \in \sigma} A(\lambda)$.

Since the set σ is compact, for a given $\varepsilon > 0$, we can find finitely many points $\lambda^{(1)}, \ldots, \lambda^{(m)}$ in σ such that for all $\lambda \in \sigma$ we have $\min\{\|\lambda - \lambda^{(i)}\|_1 : i = 1, 2, \ldots, m\} < \varepsilon/M$ where $M = \sup\{\|x\| : x \in B_i, i = 1, 2, \ldots, n\} < +\infty$. So, if $x \in \bigcup_{\lambda \in \sigma} A(\lambda)$, $x = \sum_i \lambda_i x_i$, $\lambda_i \geq 0$, $\sum_i \lambda_i = 1$ there exists $j \in \{1, 2, \ldots, m\}$ such that $\sum_{i=1}^{n} |\lambda_i - \lambda_i^j| < \varepsilon/M$. If we call $\bar{x} = \sum_{i=1}^{n} \lambda_i^j x_i$, then $\|x - \bar{x}\| \leq \sum_{i=1}^{n} |\lambda_i - \lambda_i^j| \|x_i\| < \varepsilon$ and therefore

$$\text{co}(B) \subset \bigcup_{i=1}^{m} (A(\lambda^{(i)}) + \varepsilon \overline{B}(0, 1)).$$

Hence,

$$\alpha(\text{ co}(B)) \leq \max_i\{\alpha(A(\lambda^{(i)})) + \alpha(\varepsilon\overline{B}(0,1))\} \leq \alpha(B) + \varepsilon + 2\varepsilon$$

and bearing in mind that ε was chosen arbitrarily, we obtain $\alpha(\text{ co}(B)) \leq \alpha(B)$.
□

THEOREM 2.5. *Let $B(0,1)$ be the unit ball in a Banach space X. Then $\alpha(B(0,1))$
$= \chi(B(0,1)) = 0$ if X is finite dimensional, and $\alpha(B(0,1)) = 2$, $\chi(B(0,1)) = 1$
otherwise.*

Proof. If X is a finite dimensional Banach space, the result follows from the regularity of the MNCs α and χ.

It remains to consider the infinite dimensional case. We first prove the result
for χ. Obviously $\chi(B(0,1)) \leq 1$. Suppose $\chi(B(0,1)) = r < 1$. Let us choose $\varepsilon > 0$
such that $r + \varepsilon < 1$. The there exist $x_1, x_2, \ldots, x_m$ in X such that

$$B(0,1) \subset \bigcup_{k=1}^{m} B\left(x_k, (r+\varepsilon)\right) = \bigcup_{k=1}^{m} (x_k + (r+\varepsilon)B(0,1)).$$

From the properties of the MNC χ it follows that

$$r = \chi(B(0,1)) \leq (r+\varepsilon)\chi(B(0,1)) = r(r+\varepsilon)$$

and this implies $r = 0$. So $\chi(B(0,1)) = 0$ and hence $B(0,1)$ is precompact. This
contradicts the infinite dimensionality of the space X. Therefore, $\chi(B(0,1)) = 1$.

To prove the result for α we make use of the *Borsuk-Lyusternik-Shnirel'man
theorem on antipodes*(see [Kr, page 100, Theorem 2.6]):

"If $S_n(0,1)$ is the unit sphere in an n-dimensional normed space and A_k
$(k = 1, \ldots, n)$ is a cover of $S_n(0,1)$ by closed subsets of that space, then at least one
of the sets A_k contains a pair of diametrically opposite points, that is, $\text{diam}(A_k) \geq$
$\text{diam}(S_n(0,1))$".

Since $\text{diam}(B(0,1)) = 2$, it is obvious that $\alpha(B(0,1)) \leq 2$. Suppose
$\alpha(B(0,1)) < 2$. Then we can find a finite number of closed subsets $\{B_1, B_2, \ldots, B_n\}$
of X with $\text{diam}(B_k) < 2$ for all $k = 1, \ldots, n$ such that $B(0,1) \subset \bigcup_{k=1}^{n} B_k$. Now,
taking the section of $B(0,1)$ with an arbitrary n-dimensional subspace X_n and
setting $A_k = B_k \cap X_n$, we obtain a contradiction with the theorem on antipodes.
□

Since α and χ are invariant under passage to the convex hull, we obtain the
following corollary:

COROLLARY 2.6. *Let $S(0,1)$ be the unit sphere in a Banach space X. Then
$\alpha(S(0,1)) = \chi(S(0,1)) = 0$ if X is finite dimensional, and $\alpha(S(0,1)) = 2$,
$\chi(S(0,1)) = 1$ otherwise.*

THEOREM 2.7. *The Kuratowski and Hausdorff MNCs are related by the inequalities*

$$\chi(B) \leq \alpha(B) \leq 2\chi(B).$$

In the class of all infinite dimensional Banach spaces these inequalities are the best possible.

Proof. The inequalities are obvious from the definitions of α and χ. The sharpness of the second inequality follows from theorem 2.5. The following example shows that the first inequality is also sharp. Let $B = \{e_k : k \geq 1\}$ the set of standard basis vectors in c_0. Since for all $i \neq j$, $\|e_i - e_j\| = 1$, we have $\alpha(B) = 1$. On the other hand, $\chi(B) = 1$ because the distance from any infinite subset of B to any element of c_0 cannot be smaller than 1. □

Remark 2.8. Though in general α and χ are different MNCs, in some Banach spaces we can find a direct relation between them.

Example 2: Let ℓ^∞ be the space of all real bounded sequences with the supremum norm, and let A be a bounded set in ℓ^∞. Then $\alpha(A) = 2\chi(A)$.

Indeed, we know that $\alpha(A) \leq 2\chi(A)$ in every metric space. Let ε be an arbitrary positive number and $A_1, A_2, \dots, A_r$ sets in ℓ^∞ such that A is contained in $\bigcup_{i=1}^{r} A_i$ and $\operatorname{diam}(A_i) \leq \alpha(A) + \varepsilon$. For each $k \in \mathbb{N}$ let $\alpha_i^k = \inf\{x^k : x = (x^j)_{j\in\mathbb{N}} \in A_i\}$, $\beta_i^k = \sup\{x^k : x = (x^j)_{j\in\mathbb{N}} \in A_i\}$, $c_i^k = \frac{\alpha_i^k + \beta_i^k}{2}$, $c_i = (c_i^j)_{j\in\mathbb{N}}$ and $B_i = B(c_i, \frac{\alpha(A)+\varepsilon}{2})$. It is easy to check that $A_i \subset B_i$. Thus $\chi(A) \leq \frac{\alpha(A)+\varepsilon}{2}$ and letting $\varepsilon \to 0$ we obtain $2\chi(A) \leq \alpha(A)$ and the proof is complete.

In order to obtain a useful property of the Hausdorff MNC [Ba1] we shall need the following lemma.

LEMMA 2.9. *Let A, B and C be given subsets of a Banach space X. Suppose that B is convex and closed, C is bounded and $A + C \subset B + C$. Then $A \subset B$.*

Proof. Let a be any element of A. We shall show that a must belong to B. We know that given any $c_1 \in C$ we have $a + c_1 \in B + C$, that is, there exist $b_1 \in B$ and $c_2 \in C$ such that $a + c_1 = b_1 + c_2$. For the same reason, since c_2 is in C, there exist $b_2 \in B$ and $c_3 \in C$ such that $a + c_2 = b_2 + c_3$.

Repeating the procedure indefinitely and summing the first n equations obtained, we get:

$$na + \sum_{i=1}^{n} c_i = \sum_{i=1}^{n} b_i + \sum_{i=2}^{n+1} c_i$$

or

$$a = \frac{1}{n}\sum_{i=1}^{n} b_i + \frac{c_{n+1}}{n} - \frac{c_1}{n}.$$

Since B is convex, we have that $d_n = \frac{1}{n}\sum_{i=1}^{n} b_i$ belongs to B for all $n \in \mathbb{N}$. Moreover, both c_1/n and c_{n+1}/n tend to the origin because C is bounded. So we can conclude that d_n converges to a. Since B is closed we have $a \in B$ as required. $\qquad\square$

THEOREM 2.10. $\chi(B(A,r)) = \chi(A) + r$, where $B(A,r) = \bigcup_{x \in A} B(x,r)$.

Proof. Since $B(A,r) = A + rB(0,1)$ it follows from the properties of the function χ that $\chi(B(A,r)) \le \chi(A) + r$. In order to prove the converse inequality, we notice that from the definition of χ we have, for any number $r_1 > \chi(A + rB(0,1))$, that there exists a finite set H such that $A + rB(0,1) \subset H + r_1 B(0,1)$. Thus

$$A + rB(0,1) \subset \overline{\text{co}}(H) + (r_1 - r)B(0,1) + rB(0,1).$$

Since the set $\overline{\text{co}}(H) + (r_1 - r)B(0,1)$ is closed and convex, Lemma 2.9 gives that $A \subset \overline{\text{co}}(H) + (r_1 - r)B(0,1)$, and so $\chi(A) \le \chi(H) + r_1 - r = r_1 - r$. Hence $\chi(A) + r \le r_1$, and since r_1 is an arbitrary number greater than $\chi(A + rB(0,1))$, we obtain $\chi(A) + r \le \chi(A + rB(0,1)) = \chi(B(A,r))$. This inequality completes the proof. $\qquad\square$

Finally, we are going to prove a *generalized Arzela-Ascoli theorem* using the Kuratowski measure of noncompactness [Ab].

THEOREM 2.11. *Let X be a Banach space, $D \subset \mathbb{R}^n$ compact and $B \subset \mathcal{C}(D,X)$ a bounded and equicontinuous set. Then $\alpha(B) = \sup_{t \in D} \alpha(\{x(t) : x \in B\})$.*

Proof. Let $\mu > \alpha(B)$. Then there exist finitely many sets $M_1, M_2, \ldots, M_p$, such that for every $i = 1, 2, \ldots, p$ we have $M_i \subset \mathcal{C}(D,X)$, $\text{diam}(M_i) \le \mu$ and $B \subset \bigcup_{i=1}^{p} M_i$.

Hence, for every $t \in D$ we have that $\{x(t) : x \in B\} \subset \bigcup_{i=1}^{p}\{x(t) : x \in M_i\}$ and

$$\text{diam}(\{x(t) : x \in M_i\}) = \sup_{x,x' \in M_i}\{\|x(t) - x'(t)\|\} \le \text{diam}(M_i) \le \mu.$$

Therefore $\alpha(\{x(t) : x \in B\}) \le \mu$ for every $t \in D$ and thus

$$\sup_{t \in D}\{\alpha(\{x(t) : x \in B\})\} \le \alpha(B).$$

Let us prove now the converse inequality. Since B is equicontinuous and D is a compact set, given $\varepsilon > 0$ we can find finitely many points $t_1, t_2, \ldots, t_p$ in D such that $\{x(t) : x \in B\} \subset \bigcup_{i=1}^{p}(\{x(t_i) : x \in B\} + B(0,\varepsilon))$ for any $t \in D$.

Moreover, if $\mu > \sup_{t \in D}\{\alpha(\{x(t) : x \in B\})\}$ we can find finitely many sets $M_1, M_2, \ldots, M_h$ such that $\text{diam}(M_j) \le \mu$ and $\bigcup_{i=1}^{p}\{x(t_i) : x \in B\} \subset \bigcup_{j=1}^{h} M_j$. Since B is the union of finitely many sets $\{x \in B : x(t_1) \in M_{j_1}, \ldots, x(t_p) \in M_{j_p}\}$ and these sets have a diameter which is less than $\mu + 2\varepsilon$, we obtain that $\alpha(B) \le \mu + 2\varepsilon$ and the result is reached. $\qquad\square$

3. The separation measure of noncompactness

In this section we introduce and study another MNC which is useful in applications.

Let (X, d) be a complete metric space. The set B is said to be *r-separated* if $d(x, y) \geq r$ for all $x, y \in B$, $x \neq y$. The set B will be called an *r-separation* of X.

DEFINITION 3.1. *Let $\mathcal{B}$ be the family of bounded sets on a complete metric space (X, d). For every $B \in \mathcal{B}$ we define*
$$\beta(B) = \sup\{r > 0 : B \text{ has an infinite } r\text{-separation}\}$$
or equivalently,
$$\beta(B) = \inf\{r > 0 : B \text{ does not have an infinite } r\text{-separation}\}.$$

Remark 3.2. It is not difficult to prove that β is a MNC. Moreover, if X is a Banach space, then β is semi-homogeneous, Lipschitzian with constant $L_\beta = 2$, continuous and invariant under translations. Furthermore, it is easy to check that the MNCs α, χ and β are related by the inequalities

$$\chi(B) \leq \beta(B) \leq \alpha(B) \leq 2\chi(B)$$

for all $B \in \mathcal{B}$.

Less obvious properties of the MNC β are established in the next two theorems. Although the following lemma is a straightforward consequence of Ramsey's Theorem [Ba, page 392], we shall give a direct proof for the reader who is not familiar with such combinatorial results.

LEMMA 3.3. *Let (X, d) be a metric space, B a bounded subset of X and $\{x_n\}$ a sequence in B. For any $\varepsilon > 0$, there exists a subsequence $\{x_{n_k}\}$ of $\{x_n\}$ such that $d(x_{n_j}, x_{n_k}) < \beta(B) + \varepsilon$ for every $k, j \in \mathbb{N}$.*

Proof. By Zorn's lemma we can find a maximal $(\beta(B) + \varepsilon)$−separated subset F_1 of $\{x_n\}$. The definition of $\beta(B)$ implies that F_1 is a finite set. Thus for some $n_1 \in \mathbb{N}$ the ball $B(x_{n_1}, \beta(B) + \varepsilon)$ contains an infinite subset A_1 of $\{x_n : n > n_1\}$. Suppose that x_{n_j} and A_j for $j = 1, 2, \ldots, m$ are defined with the conditions $n_{j-1} < n_j$, $x_{n_j} \in A_{j-1}$, $A_{j-1} \subset A_j$ and A_j is an infinite subset of $B(x_{n_j}, \beta(B) + \varepsilon) \cap \{x_n : n > n_j\}$. Again by Zorn's lemma we can find a maximal $(\beta(B) + \varepsilon)$-separated subset F_m of A_m which must be a finite set. For some point $x_{n_{m+1}} \in F_m$, the ball $B(x_{n_{m+1}}, \beta(B) + \varepsilon)$ contains an infinite subset $A_{m+1} \subset A_m \cap \{x_n : n > n_m\}$. Thus $A_{m+1} \subset A_m$, $n_{m+1} > n_m$ and $x_{n_{m+1}} \in A_m$. The sequence $\{x_{n_k}\}$ satisfies the condition $d(x_{n_j}, x_{n_k}) < \beta(B) + \varepsilon$ for every $k, j \in \mathbb{N}$. $\qquad\square$

THEOREM 3.4. *The MNC β is algebraically semi-additive.*

Proof. Let B_1 and B_2 be arbitrary bounded subsets of a Banach space X. Fix any $\varepsilon > 0$ and let $\{z_n\}$ be a $(\beta(B_1 + B_2) - \varepsilon)$–separated sequence in $B_1 + B_2$. We write $z_n = x_n + y_n$, where $x_n \in B_1$ and $y_n \in B_2$. By Lemma 3.3 we can extract a subsequence $\{x_{n_k}\}$ of $\{x_n\}$ such that $\|x_{n_k} - x_{n_j}\| < \beta(B_1) + \varepsilon$ for every $k, j \in \mathbb{N}$. Since

$$\beta(B_1 + B_2) - \varepsilon \leq \|z_n - z_m\| \leq \|x_n - x_m\| + \|y_n - y_m\|$$

for all $n \neq m$, we have

$$\beta(B_1 + B_2) - \beta(B_1) - 2\varepsilon \leq \|y_{n_k} - y_{n_j}\|$$

for all $k \neq j$. Thus

$$\beta(B_1 + B_2) - \beta(B_1) - 2\varepsilon \leq \beta(B_2)$$

and the arbitrariness of ε implies the required inequality $\beta(B_1 + B_2) \leq \beta(B_1) + \beta(B_2)$. $\square$

LEMMA 3.5. *Let X be a Banach space and B a bounded subset of X. For any nonnegative integer N,*

$$\beta(B) = \beta(\, co_N(B))$$

where $co_N(B)$ denotes the set of all convex combinations of at most N elements of B.

Proof. Obviously $\beta(B) \leq \beta(\, co_N(B))$. Let us see the opposite inequality. We use the representation

$$co_N(B) = \bigcup_{\lambda \in \sigma} (\lambda_1 B + \lambda_2 B + \cdots + \lambda_N B)$$

where $\sigma = \left\{ \lambda = (\lambda_1, \ldots, \lambda_N) \in \mathbb{R}^N : \lambda_i \geq 0 \ \forall i = 1, \ldots, N, \ \sum_{i=1}^N \lambda_i = 1 \right\}$. As the set σ is precompact, given any $\varepsilon > 0$ there exist $\lambda^{(1)}, \ldots, \lambda^{(m)} \in \sigma$ such that for all $\lambda \in \sigma$ there is $i \in \{1, \ldots, m\}$ with $\|\lambda - \lambda^{(i)}\|_1 < \varepsilon/M$, where $M = \sup\{\|x\| : x \in B\}$.

Let us denote by σ_ε the set $\{\lambda^{(1)}, \ldots, \lambda^{(m)}\}$ and let $x \in co_N(B)$. Hence $x = \sum_{j=1}^N \lambda_j x_j$ with $\lambda = (\lambda_1, \ldots, \lambda_N) \in \sigma$. Let $y = \sum_{j=1}^N \lambda_j^{(i)} x_j$ with $\lambda^{(i)} \in \sigma_\varepsilon$ and $\|\lambda - \lambda^{(i)}\|_1 < \varepsilon/M$.

Then $\|x - y\| = \|\sum_{j=1}^N (\lambda_j - \lambda_j^{(i)}) x_j\| \leq \sum_{j=1}^N |\lambda_j - \lambda_j^{(i)}| \|x_j\| < \varepsilon$ and so $co_N(B) \subset \bigcup_{\lambda \in \sigma_\varepsilon} (\lambda_1 B + \lambda_2 B + \cdots + \lambda_N B) + \varepsilon \overline{B}(0, 1)$. It follows that

$$\beta(\, co_N(B)) \leq \beta \left(\bigcup_{\lambda \in \sigma_\varepsilon} (\lambda_1 B + \lambda_2 B + \cdots + \lambda_N B) \right) + \varepsilon \beta(\overline{B}(0, 1))$$

$$\leq \max_{\lambda \in \sigma_\varepsilon} \beta(\lambda_1 B + \lambda_2 B + \cdots + \lambda_N B) + 2\varepsilon$$

$$\leq \max_{\lambda \in \sigma_\varepsilon} \left(\sum_{i=1}^N \lambda_i \beta(B) \right) + 2\varepsilon = \beta(B) + 2\varepsilon$$

and since $\varepsilon > 0$ is arbitrary, we obtain the result. $\square$

The invariance of β under passage to the convex hull is a rather technical result [AKPRS]. However, we include the proof because this property will be very useful in later chapters.

THEOREM 3.6. *The MNC β is invariant under passage to the convex hull.*

Proof. Obviously it suffices to prove the inequality $\beta(\text{ co}(B)) \leq \beta(B)$ for each bounded subset B of a Banach space X. Suppose by way of contradiction that $\beta(B) < \beta(\text{ co}(B))$ for some B, and numbers b and c are chosen such that $\beta(B) < b < c < \beta(\text{ co}(B))$.

To obtain a contradiction we consider certain sequences of sets, functions and numbers, described as follows. We choose in $\text{co}(B)$ an infinite r_1-separated set $\overline{Y}_1$ with $r_1 > c$. Then, there is $\overline{x} \in B$ and $\overline{Y}_{11} \subset \overline{Y}_1$ such that $\overline{Y}_{11}$ is infinite and $\|\overline{x} - y\| > r_1$ for all $y \in \overline{Y}_{11}$. Setting $y_1 = \overline{x}$, $Y_1 = \{y_1\} \cup \overline{Y}_{11}$, we get the first two objects in our construction.

Furthermore, we consider the sets $B_1^0 = B \cap \overline{B}(y_1, b)$, $B_1^1 = B \setminus B_1^0$, and notice that any $y \in \text{co}(B)$ can be represented as $y = (1 - \mu_1)u_1^0 + \mu_1 u_1^1$ with $u_1^i \in \text{co}(B_1^i)$ $(i = 0, 1)$ and $\mu_1 \in [0, 1]$. Making this representation for every fixed y, two functions $\mu_1 = \mu_1(y)$ and $u_1^i = u_1^i(y)$ are defined. The set $\{u_1^i : y \in Y_1\}$ is denoted by U_1^i and we define a binary indicator a_1 as follows: $a_1 = 1$ if $\beta(U_1^1) > c$ and $a_1 = 0$ otherwise.

In the second case, when $a_1 = 0$, one necessarily has $\beta(U_1^0) > c$. This follows from the inclusion $Y_1 \subset \text{co}_2(U_1^0 \cup U_1^1)$, Lemma 3.5, the monotonicity and semi-additivity of the function β, and the fact $\beta(Y_1) > c$. Therefore, in either of the two cases, $\beta(U_1^{a_1}) > c$.

Proceeding in an analogous manner we construct objects Y_n, y_n, B_n^i, μ_n, u_n^i, U_n^i $(i = 0, 1)$ and a_n for each positive integer n.

Since $\beta(U_n^{a_n}) > c$ and $U_n^{a_n} \subset \text{co}(B_n^{a_n})$, there exists an infinite r_{n+1}-separated set $Y_{n+1} \subset \text{co}(B_n^{a_n})$ with $r_{n+1} > c$. Let y_{n+1} be such that $y_{n+1} \in Y_{n+1} \cap B_n^{a_n}$ and $Y_{n+1} \setminus y_{n+1} \subset U_n^{a_n}$.

We set $B_{n+1}^0 = B_n^{a_n} \cap \overline{B}(y_{n+1}, b)$, $B_{n+1}^1 = B_n^{a_n} \setminus B_{n+1}^0$ and define, as in the first step, the functions $\mu_{n+1} = \mu_{n+1}(u_n^{a_n})$ and $u_{n+1}^i = u_{n+1}^i(u_n^{a_n})$ for each $u_n^{a_n} \in \text{co}(B_n^{a_n})$ such that $u_n^{a_n} = (1 - \mu_{n+1})u_{n+1}^0 + \mu_{n+1}u_{n+1}^1$, and analogously the sets $U_{n+1}^i = \{u_{n+1}^i : u_n^{a_n} \in Y_{n+1}\}$ $(i = 0, 1)$. Finally, we write $a_{n+1} = 1$ if $\beta(U_{n+1}^1) > c$ and $a_{n+1} = 0$ otherwise. Throughout the construction $\beta(U_{n+1}^{a_{n+1}}) > c$.

Notice that if $a_n = 1$ and $m \geq n + 1$, then $\|y_m - y_n\| > b$. In fact, by construction, $y_m \in B_m^0 \subset B_n^{a_n} = B_n^1$, that is, $y_m \notin \overline{B}(y_n, b)$ and so $\|y_m - y_n\| > b$. This immediately implies that the set $\{y_n : a_n = 1\}$ is b-separated in B and since $\beta(B) < b$ it follows that $\{n \in \mathbb{N} : a_n = 1\}$ is a finite set. Therefore, there is $k \in \mathbb{N}$ such that $a_n = 0$ for all $n \geq k$ and hence $Y_n \setminus \{y_n\} \subset U_{n-1}^0$ for each $n > k$.

Let $m > k$. Since u_{n+1}^i is a function of $u_n^{a_n}$ and $a_n = 0$ for all $n \geq k$, u_m^0 is a function of u_k^0. Consider the set $Y_{k+1}^m = \{u_k^0 \in Y_{k+1} : u_m^0 \in Y_m\}$, which, as Y_{k+1}, is an infinite r_{k+1}-separated set, and so verifies $\beta(Y_{k+1}^m) \geq r_{k+1} > c$.

We claim that every element $u_k^0 \in Y_{k+1}^m$ is representable in the form

$$u_k^0 = \sum_{j=1}^{m-k} \delta_j u_{k+j}^1 + \delta u_m^0 \tag{1}$$

where $\delta_j \geq 0$, $\delta = \prod_{j=1}^{m-k}(1 - \mu_{k+j})$, $\sum_{j=1}^{m-k} \delta_j + \delta = 1$. Indeed, for $m = k+1$ this is precisely the representation $u_k^0 = (1 - \mu_{k+1})u_{k+1}^0 + \mu_{k+1}u_{k+1}^1$, and the step from m to $m+1$ is made by substituting in (1) the analogous representation for u_m^0:

$$u_k^0 = \sum_{j=1}^{m-k} \delta_j u_{k+j}^1 + \delta[(1 - \mu_{m+1})u_{m+1}^0 + \mu_{m+1}u_{m+1}^1]$$

$$= \sum_{j=1}^{m-k} \delta_j u_{k+j}^1 + \delta\mu_{m+1}u_{m+1}^1 + \delta(1 - \mu_{m+1})u_{m+1}^0.$$

This is precisely a representation of the required form. However, it is convenient to write (1) in the form

$$u_k^0 = \sum_{j=1}^{m-k-1} \delta_j u_{k+j}^1 + (\delta_{m-k} + \delta)u_m^1 + \delta(u_m^0 - u_m^1).$$

From this equality it follows that

$$Y_{k+1}^m \subset \mathrm{co}_{m-k}\left(\bigcup_{j=1}^{m-k} U_{k+j}^1\right) + \delta\overline{B}(0,d)$$

where d is the diameter of $\mathrm{co}(B)$. We next show that the μ_i admit the bound $\mu_i \geq p = \frac{c-b}{d-b} > 0$. Indeed, the bound $\mu_i \geq p = \frac{c-b}{d-b}$ follows from the relation

$$u_{i-1}^{a_{i-1}} - y_i = (1 - \mu_i)(u_i^0 - y_i) + \mu_i(u_i^1 - y_i)$$

and the inequalities $\|u_{i-1}^{a_{i-1}} - y_i\| \geq r_i > c$ bearing in mind that $u_{i-1}^{a_{i-1}} \in Y_i$, $\|u_i^0 - y_i\| \leq b$ and $\|u_i^1 - y_i\| \leq d$.

Since $\delta = \prod_{j=1}^{m-k}(1 - \mu_{k+j})$ and $\mu_{k+j} \in [0,1]$, $\mu_{k+j} \geq p > 0$, it follows that δ can be made arbitrarily small by taking a sufficiently large m. Thus, we obtain $\beta(\mathrm{co}_{m-k}(\cup_{j=1}^{m-k}U_{k+j}^1)) > c$ for a large m and so, from the previous lemma, we conclude that $\beta(\cup_{j=1}^{m-k}U_{k+j}^1) > c$. From the semi-additivity of β, we can now conclude that $\beta(U_{k+j}^1) > c$ for certain $j \geq 1$, in contradiction to $a_{j+k} = 0$. Thus the proof is complete. $\square$

Remark 3.7. A completely different proof of this result can be found in [A].

To conclude the study of the MNC β, we are now going to show a very important difference between this measure and the Kuratowski and Hausdorff's measures of noncompactness. The difference is the following: the β-measure of the unit ball is not the same for all Banach spaces. In order to prove this fact, we are going to calculate the value of $\beta(U_p)$, where U_p is the closed unit ball in $L^p(\Omega)$.

To prove the result we shall use the following lemma (which can be found in [WW, page 79, Theorem 15.1]).

LEMMA 3.8. *Let* (Ω, Σ, μ) *be a σ-finite measure space and $L^p(\Omega)$ $(1 \le p < +\infty)$ the Lebesgue space of all real valued measurable functions x such that $|x|^p$ is integrable. Let $x_1, x_2, \ldots, x_n$ be vectors in $L^p(\Omega)$ and $t_1, t_2, \ldots, t_n$ nonnegative real numbers such that $\sum_{j=1}^n t_j = 1$. Let $\gamma = \max\{1 - t_j : 1 \le j \le n\}$. Then the following inequalities hold:*

$$(a) \ \sum_{j,k=1}^n t_j t_k \|x_j - x_k\|^{2\alpha} \le 2\gamma^{2-2\alpha} \sum_{j=1}^n t_j \|x_j\|^{2\alpha} \qquad 1 \le 2\alpha \le p, \ 1 \le p \le 2$$

$$(b) \ \sum_{j,k=1}^n t_j t_k \|x_j - x_k\|^{2\alpha} \le 2\gamma^{2-2\alpha} \sum_{j=1}^n t_j \|x_j\|^{2\alpha} \qquad 1 \le 2\alpha \le q, \ 2 \le p < +\infty$$

$$(c) \ \gamma^{\beta-2} \sum_{j,k=1}^n t_j t_k \|x_j - x_k\|^{\beta} \ge 2\sum_{j=1}^n t_j \|x_j - \sum_{k=1}^n t_k x_k\|^{\beta} \qquad q \le \beta, \ 1 < p \le 2$$

$$(d) \ \gamma^{\beta-2} \sum_{j,k=1}^n t_j t_k \|x_j - x_k\|^{\beta} \ge 2\sum_{j=1}^n t_j \|x_j - \sum_{k=1}^n t_k x_k\|^{\beta} \qquad p \le \beta, \ 2 \le p < +\infty$$

where q is the conjugate index of p, that is, $1/p + 1/q = 1$ with $q = +\infty$ when $p = 1$.

If $L^p(\Omega)$ is finite-dimensional, then $\beta(U_p) = 0$ since U_p is a compact set. So, henceforth, we are going to suppose that $L^p(\Omega)$ is infinite-dimensional.

The following lemma permits a lower bound for $\beta(U_p)$ to be obtained.

LEMMA 3.9. *Let* (Ω, Σ, μ) *be a σ-finite measure space, $1 \le p < +\infty$ and $L^p(\Omega)$ infinite dimensional. Then $\beta(U_p) \ge 2^{\frac{1}{p}}$.*

Proof. First we establish the existence in Ω of an infinite sequence $\{E_n : n \in \mathbb{N}\}$ of pairwise disjoint sets of finite and positive measure. To do this we consider a sequence $E_1, E_2, E_3, \ldots$ of pairwise disjoint atoms. If this sequence is infinite the result has been obtained. So we can suppose that the sequence is finite and let E_1, $E_2, \ldots, E_k$ be a maximal sequence of pairwise disjoint atoms.

Let $F = \Omega \setminus \cup_{n=1}^k E_n$. Then F is a measurable set which does not contain any atom and the infinite dimensionality of $L^p(\Omega)$ guarantees that $\mu(F) > 0$. Since F is not an atom, there exists a measurable subset E_{k+1} of F such that

$0 < \mu(E_{k+1}) < \mu(F)$, and now the existence of a sequence of pairwise disjoint sets of finite and positive measure follows by induction.

We set $x_{n,p} = \mu(E_n)^{-\frac{1}{p}} \chi_{E_n}$ for all $n \in \mathbb{N}$. It is easy to verify that $\|x_{n,p}\| = 1$ for all $n \in \mathbb{N}$ and that $\|x_{n,p} - x_{m,p}\| = 2^{\frac{1}{p}}$ for all $n \neq m$. So, $\beta(U_p) \geq 2^{\frac{1}{p}}$. $\qquad \square$

THEOREM 3.10. *Suppose* $1 \leq p \leq 2$ *and* $L^p(\Omega)$ *satisfies the above conditions. Then*

$$\beta(U_p) = 2^{\frac{1}{p}}.$$

Proof. What remains is to show that $\beta(U_p) \leq 2^{\frac{1}{p}}$. We shall use the inequality (a) of Lemma 3.8. Indeed:

Suppose that $x_1, x_2, \ldots, x_n$ are n vectors r-separated in U_p. We take $2\alpha = p$ and $t_j = 1/n$ in (a) of Lemma 3.8 so that $\gamma = 1 - 1/n$ and the inequality becomes

$$\sum_{j,k=1}^{n} \frac{1}{n^2} \|x_j - x_k\|^p \leq 2 \left(1 - \frac{1}{n}\right)^{2-p} \sum_{j=1}^{n} \frac{1}{n} \|x_j\|^p.$$

Since $\|x_n\| \leq 1$ for all $n \in \mathbb{N}$ and $\|x_n - x_m\| \geq r$ for all $n, m \in \mathbb{N}$, $n \neq m$, we have

$$n(n-1)\frac{1}{n^2} r^p \leq 2 \left(\frac{n-1}{n}\right)^{2-p}$$

or

$$r \leq 2^{\frac{1}{p}} \left(\frac{n-1}{n}\right)^{\frac{1-p}{p}}.$$

The inequality $\beta(U_p) \leq 2^{\frac{1}{p}}$ results from letting $n \to +\infty$. $\qquad \square$

Remark 3.11. Considering the cardinal measure in $\mathbb{N}$, we have, in particular, proved that $\beta(U_p) = 2^{\frac{1}{p}}$ for ℓ^p with $1 \leq p \leq 2$.

THEOREM 3.12. *Suppose* $2 < p < +\infty$, $1/p + 1/q = 1$ *and* $L^p(\Omega)$ *in the above conditions. Then* $2^{\frac{1}{p}} \leq \beta(U_p) \leq 2^{\frac{1}{q}}$. *Moreover, if* (Ω, Σ, μ) *is not purely atomic, we have that*

$$\beta(U_p) = 2^{\frac{1}{q}}.$$

Proof. Suppose that $x_1, x_2, \ldots, x_n$ are n vectors r-separated in U_p. Take $2\alpha = q$ and $t_j = 1/n$ in (b) of Lemma 3.8, so that $\gamma = 1 - 1/n$ and the inequality becomes

$$\sum_{j,k=1}^{n} \frac{1}{n^2} \|x_j - x_k\|^q \leq 2 \left(1 - \frac{1}{n}\right)^{2-q} \sum_{j=1}^{n} \frac{1}{n} \|x_j\|^q.$$

The conditions on $\|x_n\|$ and $\|x_n - x_m\|$ combine with this to give

$$n(n-1)\frac{1}{n^2}r^q \leq 2\left(\frac{n-1}{n}\right)^{2-q}$$

or

$$r \leq 2^{\frac{1}{q}} \left(\frac{n-1}{n}\right)^{\frac{1-q}{q}}.$$

The inequality $\beta(U_p) \leq 2^{\frac{1}{q}}$ results from letting $n \to +\infty$.

In order to obtain the equality, we must find a $2^{\frac{1}{q}}$-separated sequence in U_p. Since μ is not purely atomic, Ω contains a set E of positive and finite measure each of whose measurable subsets F_1 contains a measurable subset F_2 such that $2\mu(F_2) = \mu(F_1)$. (See [Hl, page 174, exercise 2]). As a consequence there exists, for each positive integer n, pairwise disjoint subsets $\{E_{n,j}\}_{j=1}^{2^n}$ such that $\cup_{j=1}^{2^n} E_{n,j} = E$ and

$$E_{n-1,k} = E_{n,2k-1} \cup E_{n,2k} \quad \text{and} \quad \mu(E_{n,2k-1}) = \mu(E_{n,2k}) = \frac{1}{2}\mu(E_{n-1,k})$$

for $k = 1, 2, \ldots, 2^{n-1}$. Using these sets we define *Rademacher functions* $\{y_{n,p} : n \in \mathbb{N}\}$ by the formula

$$y_{n,p} = \mu(E)^{-\frac{1}{p}} \sum_{k=1}^{2^n} (-1)^{k+1} \chi_{E_{n,k}}.$$

The members of this sequence satisfy $\|y_{n,p}\| = 1$ for every $n \in \mathbb{N}$ and $\|y_{n,p} - y_{m,p}\| = 2^{\frac{1}{q}}$ for all $n \neq m$, and hence $\beta(U_p) = 2^{\frac{1}{q}}$ and the proof is complete. $\square$

Obviously this theorem does not include the case of ℓ^p-spaces. So, we are going to prove the following

THEOREM 3.13. *If $2 < p < \infty$ and U_p is the closed unit ball in ℓ^p, then $\beta(U_p) = 2^{\frac{1}{p}}$.*

Proof. We have already proved that $\beta(U_p) \geq 2^{\frac{1}{p}}$. So, what remains to be proved is the opposite inequality.

Suppose that $\{x_n\}$ is an r-separated sequence in U_p and let x be the weak limit of the sequence (taking a subsequence if necessary). Then $x \in \ell^p$ and $\|x\| \leq 1$.

Let $\varepsilon > 0$ and fix a positive integer n. There exists a positive integer N such that

$$\sum_{j>N} |x_n^j|^p < \varepsilon^p.$$

Then

$$r^p \leq \|x_n - x_m\|^p = \sum_{j=1}^{\infty} |x_n^j - x_m^j|^p \leq \sum_{j=1}^{N} |x_n^j - x_m^j|^p + \sum_{j>N}^{\infty} |x_n^j - x_m^j|^p$$

$$\leq \sum_{j=1}^{N} |x_n^j - x_m^j|^p + (1+\varepsilon)^p$$

or

$$r^p - (1+\varepsilon)^p \leq \sum_{j=1}^{N} |x_n^j - x_m^j|^p.$$

Since this inequality is independent of m and $x_m \rightharpoonup x$, we may conclude that

$$r^p - (1+\varepsilon)^p \leq \sum_{j=1}^{N} |x_n^j - x^j|^p \leq \sum_{j=1}^{\infty} |x_n^j - x^j|^p.$$

Then letting $\varepsilon \to 0$ we obtain

$$r^p - 1 \leq \sum_{j=1}^{\infty} |x_n^j - x^j|^p$$

which holds for every n. Again let $\varepsilon > 0$ and choose N such that $\sum_{j>N} |x^j|^p < \varepsilon^p$. From this last inequality we obtain

$$r^p - 1 \leq \sum_{j=1}^{N} |x_n^j - x^j|^p + \sum_{j>N} |x_n^j - x^j|^p \leq \sum_{j=1}^{N} |x_n^j - x^j|^p + (1+\varepsilon)^p.$$

Letting $n \to +\infty$, this inequality gives

$$r^p - 1 \leq \sum_{j=1}^{N} |x^j - x^j|^p + (1+\varepsilon)^p = (1+\varepsilon)^p$$

and now making $\varepsilon \to 0$ we obtain $r \leq 2^{\frac{1}{p}}$ which implies $\beta(U_p) \leq 2^{\frac{1}{p}}$. $\qquad\square$

4. Measures of noncompactness in Banach sequence spaces

Let X be a Banach space with a Schauder basis $\{x_n : n \in \mathbb{N}\}$, that is, for all $x \in X$ there is a unique sequence of real numbers $\{t_n(x)\}$ such that $x = \sum_{n=1}^{\infty} t_n(x)x_n$.

Let us consider the operators

$$P_n : X \longrightarrow X : P_n(x) = \sum_{i=1}^{n} t_i(x)x_i,$$

$$R_n : X \longrightarrow X : R_n(x) = \sum_{i=n+1}^{\infty} t_i(x)x_i.$$

Obviously P_n and R_n are linear operators and $P_n + R_n = I$. Moreover, if we denote $|||x||| = \sup_n \|P_n(x)\|$, it is easy to verify that $|||.|||$ is a norm on X and $\|x\| \leq |||x|||$ for every $x \in X$. A simple argument shows that X is also complete with respect to $|||.|||$. By the open mapping theorem, the norms $\|.\|$ and $|||.|||$ are equivalent, and therefore there exists $c > 0$ such that $\|x\| \leq |||x||| \leq c\|x\|$ for every $x \in X$.

It follows that for every $n \in \mathbb{N}$ and $x \in X$ we obtain $\|P_n(x)\| \leq \sup_m \|P_m(x)\| = |||x||| \leq c\|x\|$, and hence P_n is a continuous operator and so R_n is a continuous operator (and P_n is also compact because it is finite-dimensional). Moreover, we obtain $\|P_n\| \leq c$ for every $n \in \mathbb{N}$, that is, the sequence $\{P_n\}$ is uniformly bounded. Hence $P = \sup_n \|P_n\| < +\infty$, $\sup_n \|R_n\| < +\infty$ and we can consider the positive real number $L = \limsup_{n\to\infty} \|R_n\|$. Finally, for any bounded set $A \subset X$ we let $\|A\| = \sup_{x \in A} \|x\|$.

In order to define a new measure of noncompactness we shall need the following theorem.

THEOREM 4.1. *Let X be a Banach space with Schauder basis $\{x_n : n \in \mathbb{N}\}$ and let K be a bounded subset of X. Then K is a precompact set if and only if $\lim_{n\to\infty} \sup_{x \in K} \|R_n(x)\| = 0$.*

Proof. Suppose that K is precompact and let $\varepsilon > 0$. Then, there exist $k_1, k_2, \ldots, k_p$ in X such that $K \subset \bigcup_{i=1}^{p} B(k_i, \varepsilon)$ with $k_i = \sum_{j=1}^{\infty} a_{ij}x_j$ for all $i = 1, 2, \ldots, p$.

Since $\lim_{n\to\infty} \|k_i - P_n k_i\| = 0$ for every $i = 1, 2, \ldots, p$, there is $N \in \mathbb{N}$ such that

$$\left\| k_i - \sum_{j=1}^{N-1} a_{ij}x_j \right\| < \varepsilon$$

for all $i = 1, 2, \ldots, p$. Furthermore given $x \in K$, $x = \sum_{j=1}^{\infty} a_j x_j$ and $n > N$ we have

$$\left\| \sum_{j=n}^{\infty} a_j x_j \right\| \leq 2P \left\| \sum_{j=N}^{\infty} a_j x_j \right\|.$$

Indeed, for all $m > n$ we can write

$$\left\| \sum_{j=n}^{m} a_j x_j \right\| = \left\| P_m \left(\sum_{j=N}^{\infty} a_j x_j \right) - P_n \left(\sum_{j=N}^{\infty} a_j x_j \right) \right\| \leq \| P_m - P_n \| \left\| \sum_{j=N}^{\infty} a_j x_j \right\|$$

$$\leq 2 \sup_{n} \| P_n \| \left\| \sum_{j=N}^{\infty} a_j x_j \right\| = 2P \left\| \sum_{j=N}^{\infty} a_j x_j \right\|$$

and the claim holds letting $m \to \infty$.

So, for every $n > N$ and $x \in K$, choosing $k_i \in K$ such that $\| x - k_i \| < \varepsilon$, we obtain

$$\| R_n(x) \| \leq 2P \left\| \sum_{j=N}^{\infty} a_j x_j \right\|$$

$$\leq 2P \left(\left\| \sum_{j=1}^{\infty} a_j x_j - k_i \right\| + \left\| k_i - \sum_{j=1}^{N-1} a_{ij} x_j \right\| + \left\| \sum_{j=1}^{N-1} a_{ij} x_j - \sum_{j=1}^{N-1} a_j x_j \right\| \right)$$

$$\leq 2P(\varepsilon + \varepsilon + \| P_{N-1}(k_i) - P_{N-1}(x) \|) \leq 2P(\varepsilon + \varepsilon + \| P_{N-1} \| \| x - k_i \|)$$

$$\leq 2P(\varepsilon + \varepsilon + P\varepsilon)$$

and so $\lim_{n \to \infty} \sup_{x \in K} \| R_n(x) \| = 0$.

Conversely, suppose that $\lim_{n \to \infty} \sup_{x \in K} \| R_n(x) \| = 0$. Then, given $\varepsilon > 0$, there exists $N \in \mathbb{N}$ such that for every $n > N$ and $x \in K$ we obtain $\| R_n(x) \| < \varepsilon/2$. Let $F = \mathrm{span}(x_1, x_2, \ldots, x_N)$ and $K_0 = \{ x \in F : \| x \| \leq P \| K \| \}$. Obviously $P_N(x) \in K_0$ for every $x \in K$. Moreover, since F is finite-dimensional, we know that K_0 is a precompact set and therefore, there exist $h_1, h_2, \ldots, h_p$ in F such that $K_0 \subset \bigcup_{j=1}^{p} B(h_j, \frac{\varepsilon}{2})$.

It follows that for every $x \in K$ there exists $i \in \{ 1, 2, \ldots, p \}$ such that

$$\| x - h_i \| \leq \| P_N(x - h_i) \| + \| R_N(x) \| < \varepsilon$$

where h_i was chosen such that $P_N(x) \in B(h_i, \frac{\varepsilon}{2})$. Hence $K \subset \bigcup_{i=1}^{p} B(h_i, \varepsilon)$ and so the set is precompact. $\qquad\square$

THEOREM 4.2. *The function*

$$\mu(B) = \limsup_{n \to \infty} \| R_n B \| = \limsup_{n \to \infty} \left[\sup_{x \in B} \| R_n(x) \| \right]$$

is a measure of noncompactness on X invariant under passage to the convex hull. Moreover, the following inequality holds for any bounded subset B of X:

$$\frac{1}{L} \mu(B) \leq \chi(B) \leq \inf_{n \in \mathbb{N}} \| R_n B \| \leq \mu(B)$$

where $L = \limsup_{n \to \infty} \| R_n \|$.

Proof. Theorem 4.1 implies that $\mu(B) = 0$ if and only if B is a precompact set. Furthermore, it is easy to prove that $\mu(B_1 \cup B_2) = \max\{\mu(B_1), \mu(B_2)\}$ and the property $\mu(\overline{B}) = \mu(B)$ follows from the continuity of R_n. Therefore, μ is a measure of noncompactness. Moreover the linearity of R_n permits us to prove that $\mu(\text{co}(B)) = \mu(B)$.

Let us now prove the above inequalities:

Since P_n is a compact operator we have $\chi(P_n B) = 0$ for every bounded subset B of X. In view of $R_n B \subset B + (-P_n B)$ and $B \subset R_n B + P_n B$, we obtain $\chi(R_n B) \le \chi(B)$ and $\chi(B) \le \chi(R_n B)$, and thus $\chi(B) = \chi(R_n B)$. Now the obvious inequality $\chi(R_n B) \le \|R_n B\|$ for all $n \in \mathbb{N}$ implies $\chi(B) \le \inf_{n \in \mathbb{N}} \|R_n B\|$.

Let us now assume that $\chi(B) = \chi(R_n B) = r$ and let $\varepsilon > 0$. There exists a finite set B_0 such that $B \subset \bigcup_{x \in B_0} B(x, r + \varepsilon)$. Let $x \in B$ and $x_0 \in B_0$ such that $\|x - x_0\| < r + \varepsilon$. We have for every $n \in \mathbb{N}$

$$\|R_n(x)\| - \|R_n(x_0)\| \le \|R_n(x - x_0)\| \le \|R_n\|\|x - x_0\| < \|R_n\|(r + \varepsilon)$$

and therefore

$$\|R_n(x)\| < \|R_n\|(r + \varepsilon) + \|R_n(x_0)\| \le \|R_n\|(r + \varepsilon) + \|R_n B_0\|.$$

Since B_0 is a finite set, and therefore compact, we have $\lim_{n \to \infty} \|R_n B_0\| = 0$ and hence

$$\limsup_{n \to \infty} \|R_n B\| \le (r + \varepsilon) \limsup_{n \to \infty} \|R_n\| + 0 = L(r + \varepsilon)$$

and since $\varepsilon > 0$ was arbitrary we obtain the first inequality. As the last inequality is trivial, the proof is complete. $\qquad \square$

Therefore, the measure μ described in this theorem is equivalent to the Hausdorff measure. If $L = 1$ we obtain $\chi \equiv \mu$. This is the case when $X = c_0$ or $X = \ell^p$ $(1 \le p < \infty)$ with the standard basis $\{e_n : n \in \mathbb{N}\}$, $e_n = (0, 0, \dots, 1^{(n}, 0, 0, \dots)$. This is also the case when the following class of Banach spaces is considered:

Example 3: Let $x \in \ell^p$, $1 \le p < +\infty$. We denote x^+ and x^- as the vectors whose components are

$$(x^+)^i = \max\{x^i, 0\} = \frac{|x^i| + x^i}{2}$$

$$(x^-)^i = \max\{-x^i, 0\} = \frac{|x^i| - x^i}{2}.$$

For any $q \in [1, +\infty)$ and for $x \in \ell^p$ we denote

$$\|x\|_{p,q} = (\|x^+\|_p^q + \|x^-\|_p^q)^{\frac{1}{q}}$$
$$\|x\|_{p,\infty} = \max\{\|x^+\|_p, \|x^-\|_p\}.$$

It is easy to check that these norms are equivalent to the usual norm in ℓ^p. The corresponding spaces will be denoted by $\ell^{p,q}$ and $\ell^{p,\infty}$ respectively. If $1 < p < +\infty$, $1 \leq q \leq +\infty$ and p^* and q^* are the conjugate indices of p and q, it is straightforward to show that $(\ell^{p,q})^*$ is isometrically isomorphic to ℓ^{p^*,q^*}. These spaces were introduced by Bynum [By1] and will be repeatedly considered throughout this book. It is obvious that $\{e_n : n \in \mathbb{N}\}$, $e_n = (0, 0, \ldots, 1^{(n)}, 0, 0, \ldots)$, is a Schauder basis for these spaces and that $L = 1$. Therefore $\chi \equiv \mu$ in $\ell^{p,q}$ and $\ell^{p,\infty}$.

Generally equality does not hold. This is shown by the following example.

Example 4: Let us consider the space c of all convergent sequences with the supremum norm. The elements $x_1 = (1, 1, \ldots, 1, \ldots)$ and $x_n = e_{n-1}$ for all $n > 1$ form a Schauder basis of this space.

It is easy to verify that $\|R_n\| = 2$ for any $n \in \mathbb{N}$. Indeed, if $x = (a^1, a^2, \ldots, a^n, \ldots) \in c$ with $\|x\| = 1$, then $|a^n| \leq 1$ for all n and thus $-1 \leq \lim_{n \to \infty} a^n \leq 1$, $-2 \leq a^n - \lim_{n \to \infty} a^n \leq 2$ and therefore $\|R_n(x)\| \leq 2$ for all $n \in \mathbb{N}$. It follows that $\|R_n\| \leq 2$ for every n, and the supremum is attained, for example, at the vector $x = -x_{n+2} + \sum_{i=n+3}^{\infty} x_i$. Thus $L = 2$ and the measure μ satisfies

$$\frac{1}{2}\mu(B) \leq \chi(B) \leq \mu(B).$$

Moreover the bounds are attained. Indeed, for $B = \overline{B}(0, 1)$ we have $\chi(B) = 1$ but $\mu(B) = 2$. On the other hand, for $B = \overline{B}(0, 1) \cap c_0$ we obtain $\chi(B) = \mu(B) = 1$.

Following the same idea as in the definition of μ, a different measure of noncompactness has been defined in [AD] replacing lim sup by lim inf. We believe that this measure is, in some sense, more appropriate than μ because it has the same properties as μ and, as we will see in Chapter X, a closer connection between it and the Hausdorff measure can be given.

THEOREM 4.3. *The function*

$$\nu(B) = \liminf_{n \to \infty} \sup_{x \in B} \|R_n(x)\|$$

is a measure of noncompactness on X which is invariant under passage to the convex hull and, moreover, the following inequality holds for any bounded subset B of X:

$$\frac{1}{L}\nu(B) \leq \chi(B) \leq \nu(B)$$

where $L = \limsup_{n \to \infty} \|R_n\|$.

Proof. Let ε be an arbitrary positive number. For any $n \in \mathbb{N}$ large enough and $k \geq n$ one has

$$\|R_k x\| = \|R_k(R_n x)\| \leq (L + \varepsilon)\|R_n x\|.$$

Thus

$$\liminf_{n \to \infty} \sup\{\|R_n x\| : x \in B\} \leq \limsup_{n \to \infty} \sup\{\|R_n x\| : x \in B\}$$

$$\leq (L + \varepsilon) \liminf_{n \to \infty} \sup\{\|R_n x\| : x \in B\}$$

for every bounded subset B of X. Hence $\nu(B) \leq \mu(B) \leq L\nu(B)$ and it follows that ν is regular because this property is satisfied by μ. The remaining properties are obvious or they can be proved as for the measure μ. $\qquad\square$

Remark 4.4. It is obvious that $\mu = \nu$ if $\lim_{n \to \infty} \sup\{\|R_n x\| : x \in B\}$ exists for every bounded subset B, for instance for $X = \ell^p$ or $X = c_0$ with the canonical basis. When X is the space c of converging sequences with the supremum norm and the canonical basis we have $\mu = \nu \neq \chi$. However $\mu \neq \nu$ in $L^p([0,1])$ when the Haar system is used as a Schauder basis for the space and $1 \leq p < +\infty$, $p \neq 2$ (see [AD]).

5. The theorem of Darbo and Sadovskiĭ and applications

DEFINITION 5.1. *If X and Y are metric spaces, ϕ and λ measures of noncompactness defined on X and Y respectively, and $T : D \subset X \to Y$ a mapping, then*

(a) *T is a (ϕ, λ)-contractive operator with constant $k > 0$ (or simply k-(ϕ, λ)-contractive) if T is continuous and verifies that for every bounded subset A of D we have $\lambda(T(A)) \leq k\phi(A)$.*

    In the particular case when $X = Y$ and $\lambda = \phi$ we simply say that T is a $k - \phi$-contractive operator.

(b) *T is a (ϕ, λ)-condensing operator with constant $k > 0$ (or simply k-(ϕ, λ)-condensing) if T is continuous and verifies that for every bounded and non-precompact subset A of D we have $\lambda(T(A)) < k\phi(A)$.*

    In the particular case when $X = Y$ and $\lambda = \phi$ we simply say that T is a $k - \phi$-condensing operator. Moreover, if $k = 1$ we say that T is a ϕ-condensing operator.

Remark 5.2.

(a) If $\phi = \alpha$, the k-α-contractive (or k-α-condensing) operators are usually called k-set-contractive (or k-set-condensing) operators.

(b) If $\phi = \chi$, the k-χ-contractive (or k-χ-condensing) operators are usually called k-ball-contractive (or k-ball-condensing) operators.

(c) Every compact operator is k-(ϕ, λ)-contractive and k-(ϕ, λ)-condensing for all $k > 0$.

(d) Every k-(ϕ, λ)-condensing operator is k-(ϕ, λ)-contractive. In Remark 5.5 we shall see that the converse is not true.

We are now going to give some easy properties of these operators.

PROPOSITION 5.3. *If X, Y and Z are metric spaces, ϕ, λ and ψ measures of noncompactness defined on X, Y and Z respectively, and $T : D \subset X \to Y$ and $S : Y \to Z$ mappings, then*

(a) *If T is k-(ϕ, λ)-contractive, then T is k'-(ϕ, λ)-condensing for every $k' > k$.*

(b) *If T is k_1-(ϕ, λ)-contractive (condensing) and S is k_2-(λ, ψ)-contractive (condensing), then $S \circ T$ is $k_1 k_2$-(ϕ, ψ)-contractive (condensing).*

(c) *If X and Y are Banach spaces, λ algebraically semi-additive, $T_1 : D \subset X \to Y$ k_1-(ϕ, λ)-contractive (condensing) and $T_2 : D \subset X \to Y$ k_2-(ϕ, λ)-contractive (condensing), then $T_1 + T_2$ is $(k_1 + k_2)$-(ϕ, λ)-contractive (condensing).*

(d) *If X and Y are Banach spaces and λ is algebraically semi-additive and semi-homogeneous, then the set of k-(ϕ, λ)-contractive (condensing) operators is convex.*

(e) *If X and Y are Banach spaces and λ is invariant under passage to the convex hull, then the set of k-(ϕ, λ)-contractive (condensing) operators is convex.*

Proof. (a), (b), (c) and (d) are very easy to verify. So, we are going to prove only (e).

Let T_1 and T_2 be two k-(ϕ, λ)-contractive operators and $t \in [0, 1]$. Consider the operator $T_t = t T_1 + (1 - t) T_2$.

Clearly, for every bounded subset B of D we have $T_t(B) \subset \text{co}(T_1(B) \cup T_2(B))$, and therefore it follows that

$$\lambda(T_t(B)) \leq \lambda(\,\text{co}(T_1(B) \cup T_2(B)) = \lambda(T_1(B) \cup T_2(B))$$
$$= \max\{\lambda(T_1(B)), \lambda(T_2(B))\}.$$

Suppose for example that $\max\{\lambda(T_1(B)), \lambda(T_2(B))\} = \lambda(T_1(B))$. Then we get that

$$\lambda(T_t(B)) \leq \lambda(T_1(B)) \leq k\phi(B).$$

Thus, we have obtained that $\lambda(T_t(B)) \leq k\phi(B)$ for every bounded subset B of D. As T_t is obviously a continuous operator, we conclude that T_t is k-(ϕ, λ)-contractive. The proof is analogous for k-(ϕ, λ)-condensing operators. $\square$

Example 5: Let X be a Banach space, $D \subset X$ and T and S two operators defined from D into X such that T is compact and S is k-contractive, that is, there is $k \in [0, 1)$ such that $\|Sx - Sy\| \leq k\|x - y\|$ for all $x, y \in D$. Then $T + S$ is a k-set contractive operator.

Proof. Let B be a bounded subset of D. From the definition of the Kuratowski measure of noncompactness, we immediately obtain that $\alpha(S(B)) \leq k\alpha(B)$. On the other hand, since T is compact, we have $\alpha(T(B)) = 0$. Therefore

$$\alpha((T + S)(B)) = \alpha(T(B) + S(B)) \leq \alpha(T(B)) + \alpha(S(B)) \leq k\alpha(B).$$

Since $T + S$ is continuous we obtain that it is a k-set-contractive operator.   □

Example 6: The following example gives us a set-condensing operator which is not k-set-contractive for any $k \in [0, 1)$.

Let $\varphi : [0, 1] \to \mathbb{R}^+$ be a strictly decreasing and continuous function with the property $\varphi(0) = 1$. Let $\overline{B}(0, 1)$ be the unit ball of an infinite dimensional Banach space X and we define a mapping $T : \overline{B}(0, 1) \to \overline{B}(0, 1)$ by $T(x) = \varphi(\|x\|)x$ for every $x \in \overline{B}(0, 1)$.

First we show that T is not k-set-contractive for any $k \in [0, 1)$. Indeed, from the properties of φ, it is easy to verify that $T(\overline{B}(0, r)) \supset B(0, \varphi(r)r)$. Thus $\alpha(T(\overline{B}(0, r))) \geq \alpha(B(0, \varphi(r)r)) = 2r\varphi(r) = \varphi(r)\alpha(\overline{B}(0, r))$. Since $\varphi(r) \to 1$ when $r \to 0$ it follows that T cannot be k-set-contractive for any $k \in [0, 1)$.

Let us now prove that T is a set-condensing mapping. Indeed, let B be a nonprecompact subset of $\overline{B}(0, 1)$. Therefore $\alpha(B) = d > 0$. We take $r \in (0, \frac{d}{2})$ and define the sets $B_1 = B \cap \overline{B}(0, r)$ and $B_2 = B \setminus \overline{B}(0, r)$. Obviously $T(B) = T(B_1) \cup T(B_2)$. Moreover, since $T(B_1) \subset \overline{co}(B_1 \cup \{0\})$ we obtain that

$$\alpha(T(B_1)) \leq \alpha(\overline{co}(B_1 \cup \{0\})) = \alpha(B_1 \cup \{0\}) = \alpha(B_1) \leq 2r < d = \alpha(B).$$

On the other hand, the function φ is supposed to be strictly decreasing and thus,

$$T(B_2) \subset \{ta : 0 \leq t \leq \varphi(r), \ a \in B_2\} \subset \overline{co}(\{0\} \cup \varphi(r)B_2)$$

which implies that

$$\alpha(T(B_2)) \leq \varphi(r)\alpha(B_2)) \leq \varphi(r)\alpha(B) < \alpha(B).$$

Hence

$$\alpha(T(B)) = \alpha(T(B_1) \cup T(B_2)) = \max\{\alpha(T(B_1)), \alpha(T(B_2))\} < \alpha(B)$$

and so T is a set-condensing mapping.

The main result of this section is the following theorem, usually called the *theorem of Darbo and Sadovskiĭ* [D, 1955], [Sa1, 1967].

THEOREM 5.4. *Let X be a Banach space and ϕ a measure of noncompactness which is invariant under passage to the convex hull. Let M be a nonempty, bounded, closed and convex subset of X and $T : M \to M$ a ϕ-condensing operator. Then T has a fixed point.*

Proof. Let us choose a point $m \in M$ and denote by Σ the class of all closed and convex subsets K of M such that $m \in K$ and $T(K) \subset K$. Also set

$$B = \bigcap_{K \in \Sigma} K, \quad C = \overline{\text{co}}(T(B) \cup \{m\}).$$

Obviously $\Sigma \neq \emptyset$ as $M \in \Sigma$ and $B \neq \emptyset$ as $m \in B$. Furthermore, it is easy to prove that $T(B) \subset B$ and so we have $T : B \rightarrow B$.

Moreover $B = C$. Indeed, since $m \in B$ and $T(B) \subset B$, it follows that $C \subset B$. This implies $T(C) \subset T(B) \subset C$ and so $C \in \Sigma$, and hence $B \subset C$.

Therefore the properties of ϕ now imply that $\phi(B) = \phi(C) = \phi(T(B) \cup \{m\}) = \max\{\phi(T(B)), \phi(\{m\})\} = \phi(T(B))$. Since T is ϕ-condensing, it follows that $\phi(B) = 0$, and so B is compact. Obviously B is also convex. Thus from the Theorem I.2.1 it follows that there is a fixed point for the mapping $T : M \rightarrow M$. $\square$

Remark 5.5. The theorem of Darbo and Sadovskiĭ fails to be true, even in Hilbert spaces, if we assume that T is a k-ϕ-contractive operator with constant $k = 1$. It can be shown by means of an old and well-known example.

Example 7: Let U_2 be the closed unit ball in ℓ^2. Define the operator $T : U_2 \rightarrow U_2$ by

$$T(x) = T(x^1, x^2, x^3, \dots) = (\sqrt{1 - \|x\|^2}, x^1, x^2, x^3, \dots).$$

Then we can write $T = D + S$ where D is the one dimensional mapping

$$D(x) = D(x^1, x^2, x^3, \dots) = (\sqrt{1 - \|x\|^2}, 0, 0, 0, \dots)$$

and S is an isometry. Hence, T is a well-defined, continuous operator and for every bounded subset B of U_2 we have $\alpha(T(B)) \leq \alpha(D(B) + S(B)) \leq \alpha(D(B)) + \alpha(S(B)) = 0 + \alpha(B)$. So, T is a k-set-contractive operator with constant $k = 1$. However, it is very easy to show that T does not have fixed points.

The above example shows that the Darbo and Sadovskiĭ theorem fails for 1-set-contractive mappings. We show, however, that in the sense of category, almost all 1-set-contractive mappings are condensing and so do have fixed points. In order to prove this, for every $k > 0$ we denote by $\Sigma_k(C)$ the complete metric space of all k-set-contractive mappings from a bounded, closed and convex subset C of a Banach space X into C, endowed with the metric of the uniform convergence.

THEOREM 5.6. *The set $\mathcal{C}$ of all condensing mappings in $\Sigma_1(C)$ is residual in $\Sigma_1(C)$.*

Proof. Without loss of generality we can assume that $0 \in C$. Let N be the set of all k-set-contractive mappings ($k \in [0,1)$) in $\Sigma_1(C)$, that is, $N = \bigcup_{k<1} \Sigma_k(C)$. It is not difficult to check that N is a dense subset of $\Sigma_1(C)$. Indeed, if $T \in \Sigma_1(C)$, the sequence $T_n = \left(\frac{n}{n+1}\right)T$ verifies that $T_n \in N$ for every $n \in \mathbb{N}$ and $\|T_n - T\| \to 0$.

For each $T \in N$ we denote the α-modulus of T by k_T, that is, $k_T = \inf\{k \geq 0 : T \in \Sigma_k(C)\}$. We define the set

$$C^* = \bigcap_{n=1}^{\infty} \bigcup_{T \in N} B\left(T, \frac{1-k_T}{2n}\right).$$

Then C^* is a dense G_δ-subset of $\Sigma_1(C)$. We claim that C^* is contained in $\mathcal{C}$. Let $S \in C^*$ and A be contained in C with $\alpha(A) > 0$. Choose a positive integer n such that $\frac{1}{n} < \alpha(A)$. Since $S \in C^*$, there exists $T \in N$ such that S is in $B\left(T, \frac{1-k_T}{2n}\right)$ which implies, using the properties of α, that

$$\alpha(S(A)) \leq \alpha(T(A)) + (1-k_T)/n < k_T\alpha(A) + (1-k_T)\alpha(A) = \alpha(A).$$

Hence S belongs to $\mathcal{C}$. $\qquad\qquad\square$

Remark 5.7. Darbo proved that T has a fixed point if T belongs to $\Sigma_k(C)$ for any $k \in [0,1)$. Sadovskiĭ proved the same property for a condensing mapping $T \in \Sigma_1(C)$. Although apparently Sadovskiĭ's theorem is only a slight generalization of Darbo's theorem, the real situation is very different. We have proved in theorem 5.6 that almost all 1-set-contractive mappings are condensing. On the other hand, we are going to prove that the set of all k-set-contractive mappings ($k \in [0,1)$), that is, mappings satisfying the assumptions of Darbo's theorem, is of the first Baire category in $\Sigma_1(C)$. This result is a consequence of the following lemma.

LEMMA 5.8. *Assume that the set C is noncompact. Let T be a mapping of C into C and $\varepsilon > 0$. Then there exists a mapping S in $\Sigma_1(C)$ such that $d(T,S) < \varepsilon$ and S is not condensing.*

Proof. Let x_0 be a fixed point of T. We assume without loss of generality that $x_0 = 0$. Since C is a convex and noncompact set, $B(0,r) \cap C$ is noncompact for every $r > 0$. Let $\delta > 0$, $\delta < \varepsilon/2$ be such that $\|T(x) - T(0)\| < \varepsilon/2$ if $\|x\| \leq \delta$. We define a mapping

$$S(x) = \begin{cases} x & \text{if } x \in B(0, \frac{\delta}{2}) \cap C \\ T(x) & \text{if } x \in C \setminus B(0, \delta) \\ t_x x + (1 - t_x)T(x) & \text{if } x \in (B(0,\delta) \cap C) \setminus B(0, \frac{\delta}{2}) \end{cases}$$

where $t_x = 2(\delta - \|x\|)/\delta$. It is easy to check that S is continuous and $d(T,S) < \varepsilon$. Furthermore S is not condensing because $B(0, \frac{\delta}{2}) \cap C$ is a noncompact and mapped by S onto itself. We claim that S belongs to $\Sigma_1(C)$. Indeed, let A be a bounded subset of C and denote $A_1 = A \cap B(0, \frac{\delta}{2})$, $A_2 = A \cap \left(B(0, \delta) \setminus B(0, \frac{\delta}{2}) \right)$, $A_3 = A \cap ((C \setminus B(0, \delta))$. Then $A = A_1 \cup A_2 \cup A_3$ and $\alpha(S(A)) = \max\{\alpha(S(A_i)) : i = 1, 2, 3\}$. Since S is the identity in A_1, one has $\alpha(S(A_1)) = \alpha(A_1) \leq \alpha(A)$. Since $S = T$ in A_3, one has $\alpha(S(A_3)) = \alpha(T(A_3)) \leq \alpha(A_3) \leq \alpha(A)$. Finally if $x \in A_2$ one has $S(x) \in \mathrm{co}(\{x, T(x)\}) \subset \mathrm{co}(A_2 \cup T(A_2))$. Thus $S(A_2)$ is contained in $\mathrm{co}(A_2 \cup T(A_2))$ which implies $\alpha(S(A_2)) \leq \max\{\alpha(A_2), \alpha(T(A_2))\} = \alpha(A_2) \leq \alpha(A)$. $\qquad\square$

COROLLARY 5.9. *Assume that the set C is noncompact. Then:*
(a) *The set $\Sigma_1(C) \setminus C$ is dense in $\Sigma_1(C)$.*
(b) *The set $N = \bigcup_{k<1} \Sigma_k(C)$ is of the first Baire category in $\Sigma_1(C)$*

Proof. (a) is obvious from Lemma 5.8. To prove (b) we write $k_n = n/(n+1)$ and note that $N = \bigcup_{n=1}^{\infty} \Sigma_{k_n}(C)$ and for every k_n the closed set $\Sigma_{k_n}(C)$ is nowhere dense by Lemma 5.8. $\qquad\square$

Finally, we are going to apply the Darbo and Sadovskiĭ theorem to solve an initial-value problem in a Banach space [Sz].

THEOREM 5.10. *Let a and b be real positive numbers, I the real interval $[t_0 - a, t_0 + a]$ and $V = \overline{B}(x_0, b)$ in a Banach space X, where $t_0 \in \mathbb{R}$ and $x_0 \in X$. If $f : I \times V \to X$ is a continuous mapping such that for some constant k we have $\alpha(f(I \times W)) \leq k\alpha(W)$ for any subset W of V, then there exists at least one solution of the initial valued problem*

$$x'(t) = f(t, x(t)), \quad x(t_0) = x_0 \tag{3}$$

defined on the interval $J = [t_0 - h, t_0 + h]$, where $0 < h \leq \min\{a, b/M, 1/k\}$ and $M = \sup\{\|f(t,x)\| : (t,x) \in I \times V\}$.

Proof. Instead of (3) we consider the equivalent integral equation

$$x(t) = x_0 + \int_{t_0}^{t} f(s, x(s)) ds. \tag{4}$$

Let us consider the Banach space $\mathcal{C}(J, X)$ and define the operator

$$T : \mathcal{C}(J, V) \to \mathcal{C}(J, V)$$

by

$$T_x(t) = x_0 + \int_{t_0}^{t} f(s, x(s)) ds$$

for all $x \in \mathcal{C}(J, V)$ and $t \in J$. It is not difficult to show that T is a well defined and continuous mapping and $\mathcal{C}(J, V)$ is a closed, bounded and convex subset of $\mathcal{C}(J, X)$. Then, the integral equation (4) becomes the operator equation

$$x = Tx, \quad x \in \mathcal{C}(J, V). \tag{5}$$

Let us see that:

(I) $T(\mathcal{C}(J,V))$ is a bounded and equicontinuous subset of $\mathcal{C}(J,X)$. Indeed, for every t, $t' \in J$ and $x \in \mathcal{C}(J,V)$ we have that

$$\|T_x(t)\| \le \|x_0\| + Mh \le \|x_0\| + b$$

and

$$\|T_x(t) - T_x(t')\| \le \int_t^{t'} \|f(s, x(s))\| ds \le M|t - t'|.$$

(II) T is a set condensing operator.

Let H be a nonprecompact subset of $\mathcal{C}(J,V)$. Since $T(H)$ is a bounded and equicontinuous subset of $\mathcal{C}(J,X)$, Theorem 2.11 implies that

$$\alpha\left(T\left(H\right)\right) = \sup_{t \in J}\left\{\alpha\left(\{Tx(t) : x \in H\}\right)\right\}$$

$$= \sup_{t \in J}\left\{\alpha\left(\left\{x_0 + \int_{t_0}^t f(s, x(s))ds : x \in H\right\}\right)\right\}.$$

Following the same argument as in the proof of Theorem I.2.8 we obtain that $x_0 + \int_{t_0}^t f(s, x(s))ds \in x_0 + (t - t_0)\overline{\text{co}}\left(\{f(s, x(s)) : s \in [t_0, t]\}\right)$ and so

$$\alpha(T(H)) \le \sup_{t \in J}\{\alpha(\{x_0 + (t - t_0)\overline{\text{co}}(\{f(s,x(s)) : s \in [t_0,t], x \in H\})\})\}$$

$$\le \sup_{0 \le \lambda \le h}\{\alpha(x_0 + \lambda \overline{\text{co}}(f(J \times H)))\} = h\alpha(f(J \times H)) \le hk\alpha(H) < \alpha(H).$$

Hence we can conclude from Theorem 5.4 that T has a fixed point $\bar{x} \in \mathcal{C}(J,V)$. This fixed point is a solution of (5), hence of (4), and finally, of (3). $\qquad\square$

Remark 5.11. If a function $f = f_1 + f_2$, where f_1 satisfies the Lipschitz condition and f_2 is a compact mapping, then f satisfies the assumptions of Theorem 5.10.

Chapter III
Minimal Sets for a
Measure of Noncompactness

The notion of a ϕ-minimal set for an MNC ϕ was introduced in [Do1] in order to study the relationships between condensing mappings for Kuratowski and Hausdorff's measures of noncompactness (see Chapter X).

Felix Hausdorff (1868–1942) was born in Breslau (Germany) on 8 November, the son of a well-to-do merchant. The family moved to Leipzig in 1871. He graduated from Leipzig University in 1891 in astronomy, completing his Habilitation in 1895 with a study on the absorption of light in the atmosphere. During this period he also wrote poems, and at least one successful play under the pseudonym of Paul Mongré. In 1899 he married Charlotte Goldschmidt; they had one daughter.

In about 1900 Hausdorff became interested in Cantor's set theory. He lectured on it to three students in the summer semester of 1901. This may have been the first lecture course on set theory anywhere in Germany; Cantor himself, in his more than 40 years at Halle, never lectured on set theory.

In 1901 Hausdorff was proposed for an associate professorship at Leipzig. The faculty vote was 22 in favor and 7 opposed. In sending this result to the Minister, who had the final decision, the Dean added a note stating that the minority had voted against the appointment because Hausdorff was of the "faith of Moses". He received the appointment.

In 1910 Hausdorff went to Bonn as an associate professor, then in 1913 to Greifswald as a professor, and finally he returned to Bonn in 1921 as a professor. In 1935 the Nazis compelled him to retire. He was still permitted to publish until 1938. On 26 January 1942, with deportation threatening, Hausdorff, his wife and her sister committed suicide.

The concept of ϕ-minimal set will also be used in Chapter V to obtain simpler forms for some moduli of noncompact convexity and to simplify several proofs.

In Section 1 we give the definition of a ϕ-minimal set and we prove that every bounded and nonprecompact set has a nonprecompact ϕ-minimal subset. Following a Ramsey's theorem type argument we use this result to prove that every sequence $\{x_n\}$ in a metric space has a subsequence $\{y_n\}$ such that $\lim_{n,m;n\neq m} d(y_n, y_m)$ exists. We also use this result to find an equivalent condition for a mapping T to be β-contractive.

Bearing in mind how minimal sets can be chosen in a metric space, we distinguish some special classes of MNCs. In Section 2 we prove that the MNCs defined in the previous chapter have a different behaviour with respect to this concept. The best situation is when every bounded set A has a ϕ-minimal subset B such that $\phi(B) = \phi(A)$. In this case we say that ϕ is strictly minimalizable. This is the case for the MNC μ defined in Banach spaces with a Schauder basis. It can also be proved that the Hausdorff measure is strictly minimalizable if X and X^* are weakly compactly generated Banach spaces.

1. ϕ-minimal sets

DEFINITION 1.1. *Let X be a metric space and $\mathcal{B}$ the family of all bounded subsets of X. An infinite subset $A \in \mathcal{B}$ is said to be minimal for the measure ϕ (or, in short, ϕ-minimal) if $\phi(A) = \phi(B)$ for every infinite subset B of A.*

Examples:
 (1) Every infinite precompact set is obviously ϕ-minimal for any MNC ϕ.
 (2) In particular, every Cauchy sequence with infinite range is ϕ-minimal for any MNC ϕ.
 (3) Every infinite subset of a ϕ-minimal set is a ϕ-minimal set.
 (4) The standard bases in ℓ^p or c_0 are minimal sets for the Kuratowski and Hausdorff measures of noncompactness.
 (5) A nonprecompact convex subset of a Banach space fails to be ϕ-minimal for any measure of noncompactness ϕ.

We are now going to prove the existence of ϕ-minimal sets in bounded sets.

THEOREM 1.2. *Let X be a bounded metric space and ϕ an MNC defined onto X. Then:*
 (a) *There is a subset A of X such that A is ϕ-minimal.*
 (b) *If X is not a precompact set, A can be chosen such that $\phi(A) > 0$.*

Proof. Let $A_0 = X$; recursively let

$$\phi_{n+1} = \inf\{\phi(A) : A \subset A_n, \ A \text{ infinite}\}$$

and let A_{n+1} be chosen to be an infinite subset of A_n with

$$\phi(A_{n+1}) < \phi_{n+1} + \frac{1}{n+1}.$$

Since A_n is an infinite set, for every n we can choose $a_n \in A_n$ such that $a_n \neq a_k$ for $k = 1, 2, \ldots, n-1$. Let A be the infinite set $\{a_n : n \in \mathbb{N}\}$. Then $A \backslash A_n$ is finite for each $n \in \mathbb{N}$. Let us see that A is ϕ-minimal. Indeed, let A' be an infinite subset of A. Since $A' \backslash A_{n-1}$ is a finite set for each $n > 1$, we have

$$\phi(A) \leq \phi(A_n) < \phi_n + \frac{1}{n} \leq \phi(A' \cap A_{n-1}) + \frac{1}{n} = \phi(A') + \frac{1}{n}.$$

Hence $\phi(A') = \phi(A)$ and A is ϕ-minimal. This argument concludes part (a).

We now assume that $\phi(A) = 0$ for every ϕ-minimal subset A of X. Let $\{x_n\}$ be a sequence in X and we assume that the set $\{x_n : n \in \mathbb{N}\}$ is infinite. Then there exists an ϕ-minimal subsequence $\{y_n\}$ of $\{x_n\}$. Since $\phi(\{y_n : n \in \mathbb{N}\}) = 0$, $\{y_n : n \in \mathbb{N}\}$ is precompact and so $\{y_n\}$ has a Cauchy subsequence. Thus every sequence in X has a Cauchy subsequence. Hence X is a precompact set and the proof is complete. $\qquad\square$

LEMMA 1.3. *Let A be an α-minimal subset of a metric space (X, d). Then, for every positive number ε, there exists an infinite subset B of A such that*

$$\alpha(A) - \varepsilon < d(x, y) < \alpha(A) + \varepsilon$$

for every $x \in B$, $y \in B$, $x \neq y$.

Proof. Without loss of generality we can assume that $A = \{x_n : n \in \mathbb{N}\}$ and that $0 < d(x_n, x_m) < \alpha(A) + \varepsilon$ for all $n \neq m$.

If $\alpha(A) = 0$ the result is obvious. Suppose then that $\alpha(A) > 0$. It suffices to prove that there are infinite points in A such that

$$d(x_n, x_m) > \alpha(A) - \varepsilon \quad \forall n \neq m. \tag{1}$$

We suppose by contradicting that every subset of A satisfying (1) is finite, and a maximal subset $\{x_{j_1}, x_{j_2}, \ldots, x_{j_N}\}$ of A verifying (1) is considered. Then there exists i, $1 \leq i \leq N$, such that the set $A_1 = \{n \in \mathbb{N} : d(x_n, x_{j_i}) \leq \alpha(A) - \varepsilon, \ n > j_i\}$ is infinite. We define $\varphi(1) = j_i$ and suppose $\varphi(k)$ is defined for all $k = 1, 2, \ldots, m-1$ with the conditions $A_k \subset \mathbb{N}$, A_k infinite, $A_k \subset A_{k-1}$, $\varphi(k) \notin A_k$ and $d(x_{\varphi(k)}, x_n) \leq \alpha(A) - \varepsilon$ for every $n \in A_k$.

Since $\{x_n : n \in A_{m-1}\} \subset A$, every subset of $\{x_n : n \in A_{m-1}\}$ verifying (1) is finite and therefore we can choose a maximal subset $\{x_{h_1}, x_{h_2}, \ldots, x_{h_M}\}$ of $\{x_n : n \in A_{m-1}\}$ satisfying (1).

Therefore we have $h_i \in A_{m-1}$, $1 \leq i \leq M$, such that the set

$$A_m = \{n \in A_{m-1} : d(x_{h_i}, x_n) \leq \alpha(A) - \varepsilon, \ n > h_i\}$$

is infinite. Let us define $\varphi(m) = h_i$ and note that $\varphi(m) < \varphi(n)$ for all $m < n$. Then we obtain an infinite sequence $\{x_{\varphi(n)} : n \in \mathbb{N}\} \subset A$ such that

$$\alpha(\{x_{\varphi(n)} : n \in \mathbb{N}\}) \leq \text{diam}(\{x_{\varphi(n)} : n \in \mathbb{N}\}) \leq \alpha(A) - \varepsilon < \alpha(A)$$

contradicting the minimality of A. Thus the proof is complete. $\qquad\square$

Remark 1.4. This lemma can also be derived from Ramsey's theorem (see [Dol, Lemma 3.4]).

THEOREM 1.5. *Let $\{x_n\}$ be a bounded sequence in a metric space (X, d). Then there is a subsequence $\{y_n\}$ of $\{x_n\}$ such that $\lim_{n,m;n\neq m} d(y_n, y_m)$ exists.*

Proof. By Theorem 1.2 and Lemma 1.3 there is a nonnegative number r such that for every $\varepsilon > 0$ there exists a subsequence $\{z_n\}$ of $\{x_n\}$ such that $r - \varepsilon \leq d(z_n, z_m) \leq r + \varepsilon$ for every $n, m \in \mathbb{N}$, $n \neq m$. Taking $\varepsilon = 1/n$, $(n = 1, 2, 3, \dots)$ and by using a diagonal argument we can obtain a subsequence $\{y_n\}$ of $\{x_n\}$ such that $r - 1/n \leq d(y_n, y_m) \leq r + 1/n$ for every $n, m \in \mathbb{N}$, $n < m$. □

Remark 1.6. As a first application of the above theorem, we are now going to prove a fixed point theorem for asymptotically k-contractive mappings.

DEFINITION 1.7. *A continuous mapping T from a metric space X into another metric space Y is said to be asymptotically k-contractive (or akc in short) if*

$$\lim_{n,m;n\neq m} d(Tx_n, Tx_m) \leq k \lim_{n,m;n\neq m} d(x_n, x_m)$$

for every sequence $\{x_n\}$ in X such that both limits exist.

In order to prove a fixed point theorem for akc-mappings we prove that such mappings are k-β-contractive.

THEOREM 1.8. *A mapping T is an akc-mapping if and only if T is a k-β-contractive operator.*

Proof. Assume that T is akc. For every bounded subset A of X and every positive number ε a sequence $\{x_n\}$ in A exists such that $d(Tx_n, Tx_m) \geq \beta(T(A)) - \varepsilon$, $n \neq m$. By Theorem 1.5 we can assume without loss of generality that $\lim_{n,m;n\neq m} d(x_n, x_m)$ and $\lim_{n,m;n\neq m} d(Tx_n, Tx_m)$ exist. Since

$$\lim_{n,m;n\neq m} d(Tx_n, Tx_m) \leq k \lim_{n,m;n\neq m} d(x_n, x_m)$$

and

$$\lim_{n,m;n\neq m} d(x_n, x_m) \leq \beta(A)$$

we have $\beta(T(A)) - \varepsilon \leq k\beta(A)$. Thus T is k-β-contractive.

Conversely, if $\{x_n\}$ is a sequence in X such that both $\lim_{n,m;n\neq m} d(x_n, x_m)$ and $\lim_{n,m;n\neq m} d(Tx_n, Tx_m)$ exist, we have

$$\beta(\{x_n : n \in \mathbb{N}\}) = \lim_{n,m;n\neq m} d(x_n, x_m)$$

and

$$\beta(\{Tx_n : n \in \mathbb{N}\}) = \lim_{n,m;n\neq m} d(Tx_n, Tx_m).$$

Since

$$\beta(\{Tx_n : n \in \mathbb{N}\}) \leq k\beta(\{x_n : n \in \mathbb{N}\})$$

we obtain that T is akc. □

By using the above theorem and Theorem II.5.4 we obtain a fixed point theorem for akc-mappings:

THEOREM 1.9. . *Let X be a Banach space, C a bounded, closed and convex subset of X and $T : C \to C$ an akc-mapping with $0 \leq k < 1$. Then T has a fixed point in C.*

2. Minimalizable measures of noncompactness

Bearing in mind how the minimal sets can be chosen in a metric space, we distinguish two special classes of MNCs in the next definition.

DEFINITION 2.1. *Let ϕ be an MNC defined onto the family $\mathcal{B}$ of bounded subsets of a metric space X. We say that:*

(1) *ϕ is a minimalizable MNC if for every infinite set $A \in \mathcal{B}$ and for all $\varepsilon > 0$, there exists $B \subset A$, B ϕ-minimal such that $\phi(B) \geq \phi(A) - \varepsilon$.*

(2) *ϕ is a strictly minimalizable MNC if for every infinite set $A \in \mathcal{B}$ there exists $B \subset A$, B ϕ-minimal such that $\phi(B) = \phi(A)$.*

Remark 2.2.

(a) Obviously, every strictly minimalizable MNC is minimalizable.

(b) The discrete MNC is strictly minimalizable in every metric space.

Indeed, if we take a precompact set A, then the set itself is ϕ-minimal, and if A is not a precompact set, then Theorem 1.2 implies that there exists a subset B of A such that B is ϕ-minimal and $\phi(B) > 0$. So, $\phi(B) = \phi(A) = 1$.

The following results will show that the MNCs defined in the previous chapter exhibit quite different behaviour so far as their minimalizability is concerned. First of all, we are going to prove that the Kuratowski MNC is not, in general, minimalizable. In order to show this result, we start with the following remark:

Let $\{x_n\}$ be a bounded sequence in a Hilbert space H. Since H is reflexive, there is a subsequence $\{y_n\}$ of $\{x_n\}$ and a vector $v \in H$ such that $\{y_n\}$ is weakly convergent to v. Taking a subsequence if necessary, we can also assume that $\lim_{n \to \infty} \|y_n - v\|$ exists.

Consider the mapping $\Phi : H \to \mathbb{R} : \Phi(z) = \limsup_{n \to \infty} \|y_n - z\|$. This map takes its unique absolute minimum in v. This result is a consequence of the following lemma due to Opial [O1]:

LEMMA 2.3. *If in a Hilbert space H the sequence $\{x_n\}$ is weakly convergent to x_0, then for any $x \neq x_0$ we have*

$$\limsup_{n \to \infty} \|x_n - x\| > \limsup_{n \to \infty} \|x_n - x_0\|. \tag{2}$$

Proof. Since every weakly convergent sequence is necessarily bounded, both limits in (2) are finite. Thus, to prove this inequality, it suffices to observe that in the equality

$$\|x_n - x\|^2 = \|x_n - x_0\|^2 + \|x_0 - x\|^2 + 2(x_n - x_0) \cdot (x_0 - x)$$

the last term tends to zero as n tends to infinity.   □

Now it is easy to deduce that $\chi(\{y_n : n \in \mathbb{N}\}) = \Phi(v)$. Indeed, since $\Phi(v) = \lim_{n \to \infty} \|y_n - v\|$ it follows that for every $\varepsilon > 0$ there exists $n_0 \in \mathbb{N}$ such that $y_n \in B(v, \Phi(v) + \varepsilon)$ for all $n \geq n_0$, and hence $\chi(\{y_n : n \in \mathbb{N}\}) \leq \Phi(v)$.

Conversely, let us suppose that $\{y_n : n \in \mathbb{N}\}$ can be covered by finitely many balls with radius $r < \Phi(v)$. Then there is a ball $B(u, r)$ containing infinitely many elements of this sequence. We continue to denote the subsequence contained in this ball by $\{y_n : n \in \mathbb{N}\}$. This sequence is weakly convergent to v, $\Phi(v)$ is still given by $\lim_{n \to \infty} \|y_n - v\|$ and the function $\Phi' : H \to \mathbb{R} : \Phi'(z) = \limsup_{n \to \infty} \|y_n - z\|$ takes its unique absolute minimum in v. Therefore we obtain $\|y_n - u\| \leq r < \Phi(v)$ for all $n \in \mathbb{N}$ and thus $\Phi'(u) = \limsup_{n \to \infty} \|y_n - u\| \leq r < \Phi(v) = \Phi'(v)$ contradicting the fact that Φ' has an absolute minimum at v.

LEMMA 2.4. *Let $\{x_n\}$ be an α-minimal sequence in a Hilbert space H. Assume that $x_n \neq x_m$ if $n \neq m$. If $\{y_n\}$, v and Φ are constructed as above, then*

$$\alpha(\{x_n : n \in \mathbb{N}\}) = \alpha(\{y_n : n \in \mathbb{N}\}) = \Phi(v)\sqrt{2} = \chi(\{y_n : n \in \mathbb{N}\})\sqrt{2}.$$

Proof. We only have to prove the equality $\alpha(\{y_n : n \in \mathbb{N}\}) = \Phi(v)\sqrt{2}$.

Let ε be an arbitrary positive number and define $\eta = \alpha(\{y_n : n \in \mathbb{N}\})$. By Lemma 1.3 there is a subsequence $\{z_n\}$ of $\{y_n\}$ such that

$$\eta - \varepsilon \leq \|z_n - z_m\| \leq \eta + \varepsilon, \; \forall n \neq m. \tag{3}$$

Since $\lim_{n \to \infty} \|y_n - v\| = \Phi(v)$, we can choose a positive integer k such that

$$|\|z_n - v\| - \Phi(v)| < \varepsilon, \; n \geq k. \tag{4}$$

Fixing $k \in \mathbb{N}$, let $n > k$ be large enough so that $|(z_n - v) \cdot (z_k - v)| < \varepsilon$. From the identity

$$\|z_k - z_n\|^2 = \|z_k - v\|^2 + \|z_n - v\|^2 - (z_k - v) \cdot (z_n - v) - (z_n - v) \cdot (z_k - v)$$

we obtain by (3) and (4)

$$2(\Phi(v) + \varepsilon)^2 + 2\varepsilon \geq (\eta - \varepsilon)^2 \text{ and } (\eta + \varepsilon)^2 \geq 2(\Phi(v) - \varepsilon)^2 - 2\varepsilon.$$

Hence, letting $\varepsilon \to 0$ we obtain $\eta = \sqrt{2}\Phi(v)$.   □

LEMMA 2.5. *Let H be an infinite dimensional Hilbert space and A an α-minimal subset of the unit sphere $S(0, 1)$. Then*

$$\alpha(A) \leq \sqrt{2} < 2 = \alpha(S(0, 1)).$$

Proof. Let $\{x_n\}$ be a sequence in A with $x_n \neq x_m$ for all $n \neq m$. By Lemma 2.4 we have $\frac{\alpha(A)}{\sqrt{2}} = \inf\{\Phi(z) : z \in H\}$, where $\Phi(z) = \limsup_{n\to\infty} \|y_n - z\|$ for a subsequence $\{y_n\}$ of $\{x_n\}$. Then $\frac{\alpha(A)}{\sqrt{2}} \leq \Phi(0) = \limsup_{n\to\infty} \|y_n\| = 1$. On the other hand we know from Corollary II.2.6 that $\alpha(S(0,1)) = 2$. $\qquad\square$

COROLLARY 2.6. *The Kuratowski MNC is not, in general, minimalizable.*

The situation for the Hausdorff MNC is different. We prove that χ is strictly minimalizable for a wide class of spaces. We recall that a Banach space X is said to be *weakly compactly generated* if it contains a weakly compact set which generates X. Obviously, reflexive and separable Banach spaces satisfy this condition. The following theorem has been proved by S. Prus [personal communication].

THEOREM 2.7. *Let X and X^* be weakly compactly generated Banach spaces. Then, the Hausdorff MNC is strictly minimalizable in X.*

Proof. Since X and X^* are weakly compactly generated Banach spaces, then X has a shrinking M-basis (see [Si2, page 716]), that is, we can find two families $\{e_t\}_{t\in T}$ and $\{e_t^*\}_{t\in T}$ in X and X^* respectively, such that $e_t^*(e_s) = \delta_{ts}$, $\overline{\operatorname{span}(e_t)} = X$ and $\overline{\operatorname{span}(e_t^*)} = X^*$.

We let $V = \operatorname{span}(\{e_t\}_{t\in T})$, $V_A = \operatorname{span}(\{e_t\}_{t\in A})$, $V' = \operatorname{span}(\{e_t^*\}_{t\in T})$, $V_A' = \operatorname{span}(\{e_t^*\}_{t\in A})$ with $A \in \mathcal{F} = \{$ finite subsets of $T\}$.

With this notation it is easy to show that if $\{x_n\}$ is a sequence in X such that $\|x_n\| \leq c$ for every $n \in \mathbb{N}$ and $x_n \in V_{A_n}$ with $A_n \cap A_m = \emptyset$ for $n \neq m$, then $\{x_n\}$ is weakly convergent to zero.

In order to prove the theorem, we start by proving the following claim:

Claim: For every $\varepsilon > 0$ and $A \in \mathcal{F}$ there exists $B \in \mathcal{F}$, $B \supset A$ such that the relationship $\|x\| \leq (1+\varepsilon)\|x+y\|$ holds for all $x \in V_A$ and $y \in V_{T\setminus B}$.

Indeed, if it were not the case there would exist $\varepsilon > 0$ and $A \in \mathcal{F}$ such that for every $B \in \mathcal{F}$, $B \supset A$ we could find two elements $x \in V_A$ and $y \in V_{T\setminus B}$ with $\|x\| > (1+\varepsilon)\|x+y\|$.

Now, let us take $B_1 = A$ and $x_1 \in V_{B_1}$, $y_1 \in V_{T\setminus B_1}$ with $\|x_1\| > (1+\varepsilon)\|x_1 + y_1\|$.

Choose $B_2 \in \mathcal{F}$ such that $B_1 \subset B_2$ and $y_1 \in V_{B_2}$, then we can get two elements $x_2 \in V_{B_1}$ and $y_2 \in V_{T\setminus B_2}$ such that $\|x_2\| > (1+\varepsilon)\|x_2 + y_2\|$.

Proceeding in an analogous manner we can construct a sequence of sets

$$A = B_1 \subset B_2 \subset \cdots \subset B_n \subset \cdots$$

and two sequences of elements $\{x_n\}$ and $\{y_n\}$ such that

$$\forall n \in \mathbb{N} \quad x_n \in V_A \quad y_n \in V_{B_{n+1}}, \quad y_n \in V_{T\setminus B_n}, \quad \|x_n\| > (1+\varepsilon)\|x_n + y_n\|.$$

Let $u_n = \dfrac{x_n}{\|x_n\|}$ and $v_n = \dfrac{y_n}{\|x_n\|}$ for every $n \in \mathbb{N}$. Then

$$\|u_n\| = 1 = \frac{\|x_n\|}{\|x_n\|} > (1+\varepsilon)\left\|\frac{x_n}{\|x_n\|} + \frac{y_n}{\|x_n\|}\right\|$$

and thus, $1 > (1+\varepsilon)\|u_n + v_n\|$.

Since $\{u_n\}$ is a bounded sequence in the finite-dimensional space V_A, there is a convergent subsequence, also denoted by $\{u_n\}$. Let $u = \lim_{n\to\infty} u_n$. We have that $\|u\| = 1$. Moreover for every $n \in \mathbb{N}$ we obtain

$$\|v_n\| \leq \|u_n + v_n\| + \|u_n\| < \frac{1}{1+\varepsilon} + 1$$

and so $\{v_n\}$ is bounded and as its elements have been chosen with disjoint support we see that $\{v_n\}$ is weakly convergent to zero. Hence $u_n + v_n$ is weakly convergent to u and so

$$\|u\| = 1 \leq \liminf_{n\to\infty} \|u_n + v_n\| \leq \frac{1}{1+\varepsilon}.$$

This contradiction proves the claim.

We can now prove the result. Let A be a bounded subset of X, $M = \sup_{x\in A} \|x\|$, $\chi(A) = r$ and $B = B(0, M + r + 2)$.

Since $\chi(A) = r$, we can find a element $x_1 \in A$ such that $\|x_1\| \geq r - 1/2$. Moreover, since $\overline{\operatorname{span}(e_t)} = X$, there exist $A_1 \in \mathcal{F}$ and $y_1 \in V_{A_1}$ such that $\|x_1 - y_1\| < 1/4$.

The claim now permits us to obtain $C_1 \in \mathcal{F}$, $C_1 \supset A_1$ such that $\|x\| \leq (1 + 1/2)\|x + y\|$ is verified $\forall x \in V_{A_1}$ and $\forall y \in V_{T\setminus C_1}$.

Let $u_1^1, u_2^1, \ldots, u_{m-1}^1$ be a $\frac{1}{4}$-net in $V_{A_1} \cap B$. Then, there is an element $x_2 \in A$ such that $\|x_2 - u_i^1\| \geq r - 1/4$ for all $i = 1, 2, \ldots, m_1$. For this x_2 there exist $D_2 \in \mathcal{F}$ and $y_2 \in V_{D_2}$ such that $\|x_2 - y_2\| < 1/8$. As above, the claim now permits us to obtain $C_2 \in \mathcal{F}$, $C_2 \supset C_1 \cup D_2 := A_2$ such that $\|x\| \leq (1 + 1/4)\|x + y\|$, for all $x \in V_{A_2}$ and $y \in V_{T\setminus C_2}$.

Proceeding by induction, we obtain the following:

(i) An increasing sequence $\{A_n\}$ of subsets of $\mathcal{F}$.

(ii) A $\frac{1}{2^{n+1}}$-net $u_1^n, u_2^n, \ldots, u_{m_n}^n$ in $V_{A_n} \cap B$.

(iii) A sequence $\{x_n\}$ in A such that $\forall i = 1, 2, \ldots, m_{n-1}$ $\|x_n - u_i^{n-1}\| \geq r - \frac{1}{2^n}$.

(iv) A sequence $\{y_n\}$ such that $y_n \in V_{A_n}$ and $\|x_n - y_n\| < \frac{1}{2^{n+1}}$ for every $n \in \mathbb{N}$.

(v) $\|x\| \leq \left(1 + \frac{1}{2^n}\right)\|x + y\|$ is verified $\forall x \in V_{A_n}$ and $\forall y \in V_{T\setminus A_{n+1}}$.

From (iii) and (iv) it follows that for all $i = 1, 2, \ldots, m_{n-1}$ we have $\|u_i^{n-1} - y_n\| \geq \|u_i^{n-1} - x_n\| - \|x_n - y_n\| \geq r - \frac{3}{2^{n+1}}$.

Set $U = \cup_{n=1}^{\infty} A_n$ and take an arbitrary element $x \in V_U \cap B$. Then there exists $n_0 \in \mathbb{N}$ such that for every $n > n_0$ we have $x \in V_{A_{n-1}} \cap B$, and so there exists $i \in \{1, 2, \ldots, m_{n-1}\}$ such that $\|x - u_i^{n-1}\| \leq \frac{1}{2^n}$. Hence $\|x - y_n\| \geq r - \frac{5}{2^{n+1}}$.

Let $z \in V$. Let us choose $Z \in \mathcal{F}$ such that $z \in V_Z$ and denote by Z' the intersection $Z \cap U$ and by $Z'' = Z \setminus U$. We can then write $z = x + u$ with $x \in V_{Z'}$ and $u \in V_{Z''}$. Thus there exists $n \in \mathbb{N}$ such that $x \in V_{A_n}$, $u \in V_{T \setminus A_{n+1}}$ and so $\|x\| \leq \left(1 + \frac{1}{2^n}\right) \|x + u\|$.

Now, we can write $z - y_n = x - y_n + u$ with $x - y_n \in V_{A_n}$, $u \in V_{T \setminus A_{n+1}}$ and so $\|x - y_n\| \leq \left(1 + \frac{1}{2^n}\right) \|z - y_n\|$. It follows that $\liminf_{n \to \infty} \|z - y_n\| \geq r$ and hence $\chi(\{y_n\}) \geq r$.

Let $B' = B(0, M+r)$ and $x \in B'$. Then for every $\gamma > 0$ there exists $z \in V \cap B$ such that $\|z - x\| < \gamma$. Therefore

$$\liminf_{n \to \infty} \|x - y_n\| \geq \liminf_{n \to \infty} \|z - y_n\| - \gamma \geq r - \gamma$$

and letting $\gamma \to 0$ we obtain $\liminf_{n \to \infty} \|x - y_n\| \geq r$. It follows that

$$\liminf_{n \to \infty} \|x - x_n\| \geq \liminf_{n \to \infty} \left(\|x - y_n\| - \frac{1}{2^{n+1}} \right) \geq r$$

and so we obtain $\chi(\{x_n\}) = r$. Moreover this result is true for any subsequence of $\{x_n\}$ and thus $\{x_n\}$ is a χ-minimal subset of A and the proof is complete. $\square$

We now give an example of a Banach space in which χ is not strictly minimalizable.

Example 6: Since every separable Banach space can be isometrically embedded in ℓ^∞ (see [B, page 115, Theorem 2]), we consider $X = \ell^\infty$ and A the unit ball of a Hilbert space $H \subset X$. Since $\alpha(A) = 2$ we have that $\chi(A) = 1$. Assume that there exists an α-minimal and χ-minimal subset B of A such that $\chi(B) = \chi(A) = 1$. By example II.2 $\alpha(B) = 2$, contradicting the fact that every α-minimal subset B of A satisfies $\alpha(B) \leq \sqrt{2}$ (see Lemma 2.5).

We now consider the MNC β. We are going to see that it behaves differently from α and χ. We need two preliminary lemmas.

LEMMA 2.8. *For every bounded subset A of a metric space (X, d) we have that*

$$\beta(A) = \sup\{\alpha(B) : B \subset A, \ B \ \alpha\text{-minimal}\}.$$

Proof. Assume that A is a bounded set. Then:

(i) $\alpha(B) \leq \beta(A)$ for every α-minimal subset B of A.

Indeed, if B is an α-minimal subset of A we know that for every $\varepsilon > 0$ there exists an infinite subset B' of B such that $d(x, y) > \alpha(B) - \varepsilon$ for all $x, y \in B'$, $x \neq y$. That means that $\beta(B') > \alpha(B) - \varepsilon$ and so $\beta(A) > \alpha(B) - \varepsilon$ and keeping in mind that $\varepsilon > 0$ is arbitrary we obtain $\beta(A) \geq \alpha(B)$.

(ii) For every $\varepsilon > 0$ there exists an α-minimal subset B of A such that $\alpha(B) \geq \beta(A) - \varepsilon$.

Indeed, from the definition of β, given $\varepsilon > 0$ there is an infinite subset B' of A such that $d(x, y) > \beta(A) - \varepsilon$ for all $x, y \in B'$, $x \neq y$. Hence, any α-minimal subset B of B' verifies the required inequality. $\square$

LEMMA 2.9. *Every α-minimal subset A of a metric space (X, d) is also β-minimal and $\alpha(A) = \beta(A)$.*

Proof. Since A is α-minimal, it follows from the previous lemma that $\beta(A) = \alpha(A)$. Moreover, if there is an infinite subset B of A such that $\beta(B) < \beta(A)$ then we obtain from the previous lemma that there exists an α-minimal subset B' of B with $\alpha(B') \leq \beta(B) < \beta(A) = \alpha(A)$ which contradicts the α-minimality of A. $\square$

Now, we can prove our main result in this section concerning the MNC β.

THEOREM 2.10. *Let (X, d) be a complete metric space. Then, the MNC β is minimalizable but, in general, not strictly minimalizable.*

Proof. Let A be a bounded subset of X.

If A is α-minimal, then $\alpha(A) = \beta(A)$ and for every infinite subset B of A we know that B is α-minimal and so $\alpha(B) = \alpha(A)$. Hence $\beta(B) = \alpha(B) = \alpha(A) = \beta(A)$.

If A is not α-minimal, as $\beta(A) = \sup\{\alpha(B) : B \subset A, \; B \; \alpha\text{-minimal }\}$, given $\varepsilon > 0$ there exists an α-minimal subset B of A such that $\alpha(B) > \beta(A) - \varepsilon$. Moreover B is β-minimal and $\alpha(B) = \beta(B)$ by the α-minimality of B. This means that β is minimalizable.

In order to prove that β is not strictly minimalizable in general and conclude the proof, we need only consider the following example:

Example 7: Let $X = \ell^{\infty}(\ell^2) \subset \ell^2 \times \ell^2 \times \cdots \times \ell^2 \times \cdots$ be the subspace formed by the vectors $(x^1, x^2, \ldots, x^n, \ldots)$ with $x^n \in \ell^2$ and $\sup_n \|x^n\|_2 < +\infty$ and consider the following norm in X:

$$\|(x^1, x^2, \ldots, x^n, \ldots)\| = \sup_n \{\|x^n\|_2\}.$$

Let us consider the bounded subset of X defined by

$$A = \left\{ u_{nm} = \left(0, 0, \ldots, \frac{m-1}{m}^{(m} e_n, 0, 0, \ldots \right) : n \in \mathbb{N}, m \in \mathbb{N} \right\}.$$

Then if $n \neq k$ we have

$$\begin{aligned}
\| u_{nm} - u_{km} \| &= \| (0, 0, \ldots, \frac{m-1}{m}^{(m} (e_n - e_k), 0, 0, \ldots) \| \\
&= \frac{m-1}{m} \| e_n - e_k \|_2 \\
&= \frac{m-1}{m} \sqrt{2}
\end{aligned}$$

and if $m < j$ then

$$\begin{aligned}
\| u_{nm} - u_{kj} \| &= \| (0, 0, \ldots, \frac{m-1}{m}^{(m} e_n, 0, \ldots, -\frac{j-1}{j}^{(j} e_k, 0, \ldots) \| \\
&= \sup \left\{ \frac{m-1}{m}, \frac{j-1}{j} \right\} \\
&= \frac{j-1}{j} < 1.
\end{aligned}$$

Therefore, we see that $\mathrm{diam}(A) \leq \sqrt{2}$, and since $\beta(A) \leq \mathrm{diam}(A)$, we can conclude that $\beta(A) \leq \sqrt{2}$. In fact $\beta(A) = \sqrt{2}$. To see this, we note that the subsets A_m of A given by $A_m = \{ u_{nm} : n \in \mathbb{N} \}$ have their elements equidistant and so they are α-minimal sets. Hence for them $\beta(A_m) = \alpha(A_m) = \frac{m-1}{m} \sqrt{2}$ and thus $\beta(A) \geq \sup \{ \alpha(A_m) : m \in \mathbb{N} \} = \sqrt{2}$.

However, there is no β-minimal subset B of A such that $\beta(B) = \beta(A)$. Indeed,

(a) If $B \subset A$ is β-minimal and contains points $x_m \in A_m$ for infinitely many values of m, then $\| x_m - x_{m'} \| < 1$ whenever $m \neq m'$ and so $\beta(B) \leq 1 < \sqrt{2} = \beta(A)$.

(b) If $B \subset A$ is β-minimal with $B \cap A_m = \emptyset$ except for a finite number of m and we denote by m_0 the maximum of these m, then we obtain

$$\beta(B) \leq \beta(A_{m_0}) = \frac{m_0 - 1}{m_0} \sqrt{2} < \sqrt{2} = \beta(A).$$

$\square$

Finally, let us study the case of the measure μ defined in Banach spaces with a Schauder basis (see Theorem II.4.2). We easily obtain the following result which is the best possible.

THEOREM 2.11. *Let X be a Banach space with a Schauder basis. Then the measure of noncompactness μ is strictly minimalizable.*

Proof. If $\mu(A) = 0$ then A is μ-minimal. So, we can suppose $\mu(A) > 0$.

Since $\mu(A) = \limsup_{n\to\infty} [\sup_{x\in A} \|R_n(x)\|]$ for every bounded subset A of X, there is a subsequence $\{R_{n_k}\}$ of $\{R_n\}$ such that $\mu(A) = \lim_{k\to\infty} [\sup_{x\in A} \|R_{n_k}(x)\|]$. From the definition of a supremum, for every $k \in \mathbb{N}$ there exists $a_k \in A$ such that $\|R_{n_k}(a_k)\| \geq \sup_{x\in A} \|R_{n_k}(x)\| - \frac{1}{k}$. Now consider the set $B = \{a_k : k \in \mathbb{N}\}$ and let $C = \{a_j : j \in \mathbb{N}\}$ a subset of B. Then

$$\mu(A) = \limsup_{n\to\infty} \left[\sup_{x\in A} \|R_n(x)\|\right] = \lim_{k\to\infty} \left[\sup_{x\in A} \|R_{n_k}(x)\|\right]$$

$$= \lim_{k\to\infty} \left[\|R_{n_k}(a_k)\| + \frac{1}{k}\right] = \lim_{k\to\infty} \|R_{n_k}(a_k)\|$$

$$= \lim_{j\to\infty} \|R_{n_j}(a_j)\| \leq \mu(C) \leq \mu(B) \leq \mu(A)$$

and so B is μ-minimal and $\mu(B) = \mu(A)$. $\qquad\square$

Chapter IV
Convexity and Smoothness

In this and the following chapters we are going to study some important metric properties in the framework of Banach spaces. We call metric properties those which are invariant under isometries, in contrast to topological properties which are invariant with respect to homeomorphisms. Schauder's fixed point theorem for continuous mappings is the most celebrated topological fixed point theorem.

> **Juliusz Schauder** (1899–1943) was born in Lwów (Poland), the son of a lawyer. He studied at the University of Lwów under Banach and Steinhaus, the founders of the Polish functional analysis school during the twenties in Lwów. Personal contact with Leon Lichtenstein awakened Schauder's interest in differential equations. After receiving a Rockefeller grant in 1932, he spent some time in Leipzig with Lichtenstein. In this period he also spent time in Paris, working together with Jean Leray. The fruit of this collaboration was the fundamental work "Topologie et équations fonctionelles" which appeared in 1934. In 1938, Schauder and Leray received the Grand Prix International of Metaxas.
>
> After the occupation of Lwów by the German army in June 1941, Schauder like all others of Jewish descent became a victim of Hitler's persecution. He was forced into hiding together with his wife and small daughter, under an assumed name. In 1943, during an extermination action, he, and shortly thereafter his wife were murdered by the Nazis. His daughter survived.

However, metric properties play an essential role in fixed point theory for nonexpansive mappings. In this chapter we are going to study the concepts of convexity and smoothness in Banach spaces.

1. Strict convexity and smoothness

The definition of strictly convex spaces, which is fundamental for the geometric theory of Banach spaces is due to Clarkson [C]. This section will be devoted to this and the dual concept of a smooth space.

DEFINITION 1.1. *A Banach space X is said to be strictly convex if whenever x and y are not collinear vectors of X, then*

$$\|x + y\| < \|x\| + \|y\|.$$

Example 1: $(\mathbb{R}^n, \|.\|_\infty)$ is not strictly convex while the spaces $(\mathbb{R}^n, \|.\|_p)$, $1 < p < \infty$, are strictly convex.

DEFINITION 1.2. *Let C be a convex set in a Banach space X. A point z in C is said to be an extreme point for C if whenever $z = tx + (1-t)y$ for some t in $(0,1)$ and some x and y in C, then $x = y$.*

We give the following characterizations of strictly convex spaces:

THEOREM 1.3. *Let X be a Banach space. The following assertions are equivalent:*

(1) *X is strictly convex.*
(2) *If $\|x\| = \|y\| = 1$ and $x \neq y$ then $\|\frac{x+y}{2}\| < 1$.*
(3) *For every p, $1 < p < +\infty$, and for all $x \neq y$ in X*

$$\left\|\frac{x+y}{2}\right\|^p < \frac{1}{2}(\|x\|^p + \|y\|^p).$$

(4) *Every x in X, with $\|x\| = 1$, is an extreme point of the closed unit ball of X.*
(5) *For each non zero $f \in X^*$ there is at most one point, x, in the closed unit ball at which f attains its norm, that is, $f(x) = \|f\|$.*

Proof.

(a) $1 \Leftrightarrow 2$

If X is strictly convex and x and y are noncollinear vectors with $\|x\| = \|y\| = 1$, then $\|(x+y)/2\| < 1$. If x, y are collinear and $x \neq y$, then $x + y = 0$ and $\|x + y\| = 0$.

Conversely let us assume that noncollinear vectors x and y exist in X with $\|x + y\| = \|x\| + \|y\|$. We can assume that $1 = \|y\| > \|x\|$. We shall prove that in this case $\left\|\frac{x}{\|x\|} + y\right\| = 1 + \|y\|$.

Let $\phi_1(t) = \|tx + y\|$ and $\phi_2(t) = t\|x\| + \|y\|$. Then $\phi_1(t)$ is a convex mapping, $\phi_2(t)$ is affine, $\phi_1(0) = \phi_2(0)$, $\phi_1(1) = \phi_2(1)$ and $\phi_1(t) \leq \phi_2(t)$ for every $t \geq 0$. So $\phi_1(t) = \phi_2(t)$. Taking $t = 1/\|x\|$ we obtain the conclusion.

(b) $1 \Leftrightarrow 3$

Let X be strictly convex. If x and y are noncollinear vectors, then $\|x + y\| < \|x\| + \|y\|$. Since t^p is a convex function for $p > 1$, we have

$$\left\|\frac{x+y}{2}\right\|^p < \left(\frac{\|x\|}{2} + \frac{\|y\|}{2}\right)^p \leq \frac{1}{2}(\|x\|^p + \|y\|^p).$$

If x and $y \in X$ are collinear vectors, $y = tx$, $t \in \mathbb{R}$, $t \neq 1$ we distinguish two cases

Case $t \geq 0$. Here,

$$\left\| \frac{x+y}{2} \right\|^p = \left\| \left(\frac{1+t}{2} \right) x \right\|^p = \left| \frac{1+t}{2} \right|^p \|x\|^p$$
$$< \frac{1}{2}(1+t^p)\|x\|^p = \frac{1}{2}(\|x\|^p + \|y\|^p).$$

Case $t \leq 0$. In this case we have

$$\left\| \frac{x+y}{2} \right\|^p = \left| \frac{1+t}{2} \right|^p \|x\|^p \leq \left(\frac{1+|t|}{2} \right)^p \|x\|^p$$
$$\leq \frac{1}{2}(\|x\|^p + \|y\|^p).$$

Conversely, suppose X satisfies (3). If $\|x\| = \|y\| = 1$ then $\|(x+y)/2\|^p < 1$. So $\|x+y\| < 2$.

(c) $1 \Leftrightarrow 4$

Suppose X is not strictly convex. Then x and y exist, $x \neq y$, in X, such that $\|x\| = \|y\| = 1$ and $\|(x+y)/2\| = 1$. Hence $z = (1/2)x + (1/2)y$ is not an extreme point and $\|z\| = 1$.

Conversely if there exists z in the unit sphere such that z is not an extreme point, then $z = \lambda_0 x + (1 - \lambda_0)y$ for some $\lambda_0 \in (0,1)$, and for some x, y with $\|x\| = \|y\| = 1$ and $x \neq y$. We shall show that in this case $\|x+y\| = 2$. Suppose this were not the case. Then $\|x+y\| < 2$ and so

$$1 = \|\lambda_0 z + (1 - \lambda_0)z\| = \|\lambda_0^2 x + \lambda_0(1-\lambda_0)(x+y) + (1-\lambda_0)^2 y\| <$$
$$< \lambda_0^2 + 2\lambda_0(1-\lambda_0) + (1-\lambda_0)^2 = 1$$

which is a contradiction.

(d) $2 \Leftrightarrow 5$

Suppose (2) is satisfied and f is an element of X^*. We may suppose without loss of generality that the functional f has norm one. If x and y are norm one vectors such that $f(x) = f(y) = 1$, then $2 = f(x+y) \leq \|x+y\| = 2$ and so by hypothesis $x = y$.

Suppose (5) is satisfied and x, y are norm one vectors of X. Then if $\|(x+y)/2\| = 1$, a norm one functional exists such that $f((x+y)/2) = 1$. Since $f(x/2) \leq 1/2$ and $f(y/2) \leq 1/2$ we may deduce that $f(x) = f(y) = 1$ and, by statement (5), $x = y$. $\qquad \square$

Example 2: Let X be a Banach space and let $1 < p < +\infty$. The space

$$\ell^p(X) = \{x = (x^i) : x^i \in X \text{ and } \sum_{i=1}^{\infty} \|x^i\|^p < \infty\}$$

is a Banach space with respect to the norm $\|(x^i)\| = (\sum_{i=1}^{\infty} \|x^i\|^p)^{\frac{1}{p}}$. Moreover $\ell^p(X)$ is strictly convex if and only if X is strictly convex. We shall prove only the last statement.

Let $\ell^p(X)$ be strictly convex and x, y be in X such that $\|x\| = \|y\| = 1$ and $x \neq y$. If we take $u = (x, 0, 0, \dots)$ and $v = (y, 0, 0, \dots)$ in $\ell^p(X)$, then $\|u\| = \|v\| = 1$. So $\|u + v\| = \|x + y\| < 2$.

Conversely, let X be strictly convex and u, v be in $\ell^p(X)$ with $\|u\| = \|v\| = 1$ and $\|u + v\| = 2$. If $u = (x^i)$ and $v = (y^i)$ by Minkowski's inequality $\|x^i\| + \|y^i\| = \|x^i + y^i\|$ for all $i \in \mathbb{N}$ which, by strict convexity of X_i, implies $x^i = y^i$ for all i, that is $u = v$.

In particular considering $X = \mathbb{R}$ we may deduce that ℓ^p spaces are strictly convex for $1 < p < \infty$.

Bearing in mind that not all Banach spaces are strictly convex, it is natural to give conditions under which a Banach space has an equivalent strictly convex norm. These spaces are called *strictly convexifiable*. In 1936 Clarkson proved that every separable Banach space is strictly convexifiable. The proof of this result which we include here is due to Klee [Kl].

LEMMA 1.4. *A Banach space X is strictly convexifiable if and only if there exists a strictly convex Banach space Y and a continuous and linear operator, $T : X \longrightarrow Y$, which is injective.*

Proof. If we define in X the equivalent norm $\|x\| = \|x\| + \|Tx\|$, it is easy to see that $(X, \| \cdot \|)$ is a strictly convex Banach space. $\qquad \square$

THEOREM 1.5. *Every separable Banach space X is strictly convexifiable.*

Proof. Let $\{x_n\}$ be a norm one dense sequence in the unit ball of X and $\{x_n^*\}$ a norm one sequence of elements of X^* such that $x_n^*(x_n) = 1$.

If for each x in X we define the operator $T : X \longrightarrow \ell^2$,

$$T(x) = \sum_{n=1}^{\infty} 2^{-n} x_n^*(x) e_n,$$

where $\{e_n\}$ is the canonic basis of ℓ^2, then T is an injective, continuous and linear operator and the result follows by lemma 1.4. $\qquad \square$

In Theorem 1.3 we have proved that strict convexity is equivalent to the condition that every supporting hyperplane of the closed unit ball of X meets unit ball in at most one point. Indeed this is equivalent to saying that for every f in X^* with norm one there exists at most one x_0 in the closed unit ball of X such that $f(x_0) = 1$. If we consider the dual version of the above condition we have the following definition.

DEFINITION 1.6. *A Banach space X is said to be smooth if, for every x in X with norm one, there is a unique f in X^* with $\|f\| = 1$ such that $f(x) = 1$.*

From the fifth statement of Theorem 1.3 it is direct the following result

THEOREM 1.7. *Let X be a Banach space, then :*
(a) *If X^* is smooth, X is strictly convex.*
(b) *If X^* is strictly convex, X is smooth.*

Remark 1.8. From example 2 and theorem 1.7 we may deduce that ℓ^p spaces are smooth for $1 < p < \infty$.

In general the converses of statements (a) and (b) are not true as we can see in the following example.

Example 3: $(\ell^1, \|\cdot\|)$ where $\|x\| = (\|x\|_1^2 + \|x\|_2^2)^{\frac{1}{2}}$ is a space isomorfic to ℓ^1 with the usual norm. The dual $(\ell^\infty, \|\cdot\|^*)$ is therefore isomorfic to ℓ^∞. We are going to prove that $(\ell^1, \|\cdot\|)$ is strictly convex but $(\ell^\infty, \|\cdot\|^*)$ is not smooth. We leave the reader to check that $(\ell^1, \|\cdot\|)$ is strictly convex and will prove that $(\ell^\infty, \|\cdot\|^*)$ is not smooth.

In the first step we prove that if $y = (1, 1, 1, 1, \dots)$ then $\|y\|^* = 1$.

If $\|x\| = 1$ then $\|x\|_1 \leq 1$. So $|y(x)| \leq \|y\|_\infty \|x\|_1 \leq 1$. Hence $\|y\|^* \leq 1$.
On the other hand if

$$x_n = (\frac{1}{n}, \frac{1}{n}, \dots, \frac{1}{n}^{n-1)}, 0, 0, \dots)$$

then $\|x_n\|^2 = 1 - 1/n < 1$ and $\lim_{n\to\infty} y(x_n) = 1$. So $\|y\|^* = 1$.

The second step is to show that if $z = (z^k) \in \ell^\infty$ and $\|z\|^* = 1$, then for all $n \in \mathbb{N}$ we have

$$\sum_{k=1}^n |z^k| \leq \sqrt{n^2 + n}.$$

To prove this step let n be a fixed natural number and $x = (x^k)$ be the element of $(\ell^1, \|\cdot\|)$, with $x^k = \operatorname{sgn}(z^k)$ if $k \leq n$ and $x^k = 0$ if $k > n$. Notice that $\|x\| = \sqrt{n^2 + n}$. Since

$$\left| z\left(\frac{x}{\|x\|}\right) \right| = \frac{\sum_{k=1}^n |z^k|}{\sqrt{n^2 + n}} \leq 1,$$

we may deduce the result.

Finally let U be a free ultrafilter on $\mathbb{N}$. On ℓ^∞ we define the functional

$$\Phi_U(x^n) = \lim_U x^n.$$

Then Φ_U is a norm one element of $(\ell^\infty, \|\cdot\|^*)^*$. Indeed, notice that

$$\Phi_U(1,1,\ldots,1,\ldots) = 1.$$

So it only remains to prove that $\Phi_U(z) \leq 1$ for all norm one vectors of $(\ell^\infty, \|\cdot\|^*)$.

It will be enough to prove that for all $\varepsilon > 0$ the set $A = \{k : |z^k| > 1 + \varepsilon\}$ is finite.

We can reorder the components of the vector z while maintaining the norm of the new vector as less or equal to 1. Indeed, let $\hat{z}$ denote the reordered vector $(z^{\psi(k)})$. If $|\hat{z}(x)| > 1$ for some $x = (x^k)$ with $\|x\| = 1$ we have

$$1 < \left|\sum_{k \in \mathbb{N}} z^{\psi(k)} x^k\right| = \left|\sum_{k \in \mathbb{N}} z^k x^{\psi^{-1}(k)}\right|$$

which is impossible because

$$\|(x^{\psi^{-1}(k)})\| = \|(x^k)\| = 1.$$

If set A were infinite, then for the previous step the inequality $\sum_{k=1}^n (1+\varepsilon) \leq \sqrt{n^2 + n}$ would hold for all $n \in \mathbb{N}$, which is a contradiction.

Since two different free ultrafilters exist on $\mathbb{N}$, we can deduce that the space $(\ell^\infty, \|\cdot\|^*)$ is not smooth.

2. k-uniform convexity

If X is a strictly convex space, for each x and y in X with $\|x\| = \|y\| = 1$ and $x \neq y$ there exists $\delta(x,y) > 0$ such that $\|(x+y)/2\| < 1 - \delta(x,y)$. A natural question is whether or not $\delta(x,y)$ is uniformly bounded from below for some $\delta > 0$ when $\|x - y\|$ is bounded from below for some $\varepsilon > 0$. The following example does not verify this property.

Example 4: The Banach space $(C[0,1], \|\cdot\|)$, where $\|x\| = \|x\|_\infty + (1/2)\|x\|_2$, is strictly convex and does not have the aforementioned property.

First we show that the space is not strictly convex.

Suppose x, y are non collinear vectors in $(C[0,1], \|\cdot\|)$. Then

$$\|x + y\| = \|x + y\|_\infty + \frac{1}{2}\|x + y\|_2$$
$$< \|x\|_\infty + \|y\|_\infty + \frac{1}{2}(\|x\|_2 + \|y\|_2)$$
$$= \|x\| + \|y\|.$$

So $(C[0,1], \|\cdot\|)$ is not strictly convex.

Now for $n \geq 2$, let f_n and g_n denote the continuous functions defined as

$$f_n(t) = \begin{cases} 1, & \text{for } t \in [0, 1/n] \\ -nt + 2, & \text{for } t \in [1/n, 2/n] \\ 0, & \text{for } t \in [2/n, 1] \end{cases}$$

and

$$g_n(t) = \begin{cases} 1, & \text{for } t \in [0, 1/2n] \\ -2nt + 2, & \text{for } t \in [1/2n, 1/n] \\ 0, & \text{for } t \in [1/n, 1] \end{cases}$$

If we consider in $(C[0, 1], \|.\|)$ the sequences of elements $\{x_n\}$ and $\{y_n\}$, where $x_n = f_n/\|f_n\|$ and $y_n = g_n/\|g_n\|$, it is easy to check that

(1) $\|x_n\| = \|y_n\| = 1$,
(2) $\lim_{n \to \infty} \|(x_n + y_n)/2\| = 1$,
(3) $\lim_{n \to \infty} \|x_n - y_n\| = 1$.

The above consideration lead us to introduce the definition of a uniformly convex space.

DEFINITION 2.1. *A Banach space X is said to be uniformly convex, UC for short, if for each $\varepsilon \in (0, 2]$ there exists $\delta > 0$ such that for $x, y \in X$ with*

$$\left.\begin{array}{l} \|x - y\| \geq \varepsilon \\ x, y \in \overline{B}(0, 1) \end{array}\right\} \Rightarrow 1 - \left\|\frac{x + y}{2}\right\| > \delta.$$

Example 5: A Hilbert spaces H is UC. This fact is a direct consequence of Parallelogram Identity. Indeed, if $x, y \in \overline{B}(0, 1)$ and $\|x - y\| \geq \varepsilon$, then

$$\left\|\frac{x + y}{2}\right\| \leq \sqrt{1 - \left(\frac{\varepsilon}{2}\right)^2}.$$

Considering $\delta = \sqrt{1 - (\varepsilon/2)^2}$ we can deduce that H is UC.

The following theorem is an useful tool when checking if a Banach space is uniformly convex.

THEOREM 2.2. *X is uniformly convex if and only if it always follows from $\|x_n\| \leq 1$, $\|y_n\| \leq 1$ and $\lim_{n \to \infty} \|(x_n + y_n)/2\| = 1$ that $\lim_{n \to \infty} \|x_n - y_n\| = 0$.*

Proof. If X is not uniform convex then there exists $\varepsilon > 0$ such that for all n in $\mathbb{N}$, there exist sequences $\{x_n\}$ and $\{y_n\}$ in the unit ball which satisfy $\|(x_n + y_n)/2\| \geq 1 - 1/n$ and $\|x_n - y_n\| > \varepsilon$. So $\lim_{n \to \infty} \|(x_n + y_n)/2\| = 1$ and $\lim_n \|x_n - y_n\| \neq 0$.

Conversely suppose $\{x_n\}$ and $\{y_n\}$ are two sequences such that $\|x_n\| \leq 1$, $\|y_n\| \leq 1$, $\lim_{n \to \infty} \|(x_n + y_n)/2\| = 1$ but $\lim_{n \to \infty} \|x_n - y_n\| \neq 0$. Then taking a subsequence of $\{x_n - y_n\}$ such that $\|x_{n_k} - y_{n_k}\| > \varepsilon$ for some $\varepsilon > 0$, it is easy to see that X is not uniformly convex. $\qquad\square$

From this theorem is easy to deduce that if the unit ball is compact then uniform convexity and strict convexity are equivalent notions.

For our purpose it will be useful to have a measure of the convexity of the unit ball. Hence we consider the following definition due to Day [Da2].

DEFINITION 2.3. *Let X be a Banach space. We define the modulus of convexity of X, $\delta_X(\varepsilon)$, as follows:*

$$\delta_X(\varepsilon) = \inf\left\{ 1 - \left\| \frac{x+y}{2} \right\| : x, y \in \overline{B}(0,1), \ \|x - y\| \geq \varepsilon \right\}.$$

Remark 2.4.

(a) The modulus of convexity is a real valued function defined from $[0,2]$ to $[0,1]$ which is continuous on $[0,2)$. The proof of this result and some other properties of $\delta_X(\varepsilon)$ can be found in [GK1, Chapter 5].

(b) A Banach space is uniformly convex if and only if $\delta_X(\varepsilon) > 0$ for $\varepsilon > 0$

Example 6: Let $p \in \mathbb{R}$, $2 \leq p < \infty$. It is well known the exact value of the modulus of convexity for the spaces ℓ^p and $L^p(\Omega)$ [B, Lemma 2.9, page 98]

$$\delta_{\ell^p}(\varepsilon) = \delta_{L^p(\Omega)}(\varepsilon) = 1 - \left(1 - \left(\frac{\varepsilon}{2} \right)^p \right)^{1/p}. \tag{1}$$

The case $1 < p < 2$ is a bit more difficult [Ha]. Suppose $X = \ell^p$ or that $X = L^p(\Omega)$. Then the modulus of convexity δ_X is calculated from the equation

$$\left(1 - \delta_X(\varepsilon) + \frac{\varepsilon}{2} \right)^p + \left(1 - \delta_X(\varepsilon) - \frac{\varepsilon}{2} \right)^p = 2. \tag{2}$$

From the relations (1) and (2) and Remark 2.4 we may deduce that for $1 < p < \infty$ the spaces ℓ^p and $L^p(\Omega)$ are uniformly convex.

Roughly speaking, we can say that uniform convexity corresponds to the idea that segments with fixed positive length not can be found as near to the unit sphere as we want. If we want to work with the possibility that for some length a suitable segment could be found, then the following concept is useful.

DEFINITION 2.5. *The characteristic of convexity of a Banach space X is the constant associated with the space given by*

$$\varepsilon_0(X) = \sup\{\varepsilon : \delta_X(\varepsilon) = 0\}.$$

Obviously a space X is uniformly convex if and only if $\varepsilon_0(X) = 0$.

Example 7: Let $\ell^{p,q}$ be the Banach spaces introduced in Example II.3. We are going to check that $\varepsilon_0(\ell^{p,\infty}) = 1$.

Assume $\delta_{\ell^p,\infty}(\varepsilon) = 0$ for some $\varepsilon \in [0,2]$. Then there are sequences $\{x_n\}$ and $\{y_n\}$ in the unit ball such that $\|x_n - y_n\|_{p,\infty} \geq \varepsilon$ and $\|x_n + y_n\|_{p,\infty} \to 2$ as $n \to \infty$. Since the sequences $\{-x_n\}$ and $\{-y_n\}$ satisfy the same property we can assume

$$\|x_n + y_n\|_{p,\infty} = \|(x_n + y_n)^+\|_p.$$

Since we can exchange $x_n - y_n$ for $y_n - x_n$ we can also assume that

$$\|x_n - y_n\|_{p,\infty} = \|(x_n - y_n)^+\|_p.$$

Note that

$$\|(x_n + y_n)^+\|_p \leq \|x_n^+ + y_n^+\|_p.$$

Indeed, if the component $((x_n + y_n)^+)^i$ is bigger than zero, then $x_n^i > 0$ or $y_n^i > 0$. We have in this case

$$((x_n + y_n)^+)^i = (x_n^i + y_n^i) \leq x_n^i = (x_n^+)^i \text{ if } y_n^i \leq 0;$$

$$((x_n + y_n)^+)^i \leq (y_n^+)^i \text{ if } x_n^i \leq 0;$$

or

$$((x_n + y_n)^+)^i = (x_n^+)^i + (y_n^+)^i \text{ if } x_n^i > 0 \text{ and } y_n^i > 0.$$

Thus we have

$$2 \geq \|x_n^+ + y_n^+\|_p \geq \|x_n + y_n\|_p \to 2.$$

Therefore the uniform convexity of ℓ^p implies $\|x_n^+ - y_n^+\|_p \to 0$. The same argument as above proves that $\|((x_n^+ - y_n^+) + (y_n^- - x_n^-))^+\|_p \leq \|(x_n^+ - y_n^+)^+ + (y_n^- - x_n^-)^+\|_p$. Thus we have

$$\varepsilon \leq \|x_n - y_n\|_{p,\infty} = \|(x_n - y_n)^+\|_p = \|((x_n^+ - y_n^+) + (y_n^- - x_n^-))^+\|_p$$

$$\leq \|(x_n^+ - y_n^+)^+ + (y_n^- - x_n^-)^+\|_p \leq \|(x_n^+ - y_n^+)^+\|_p + \|(y_n^- - x_n^-)^+\|_p.$$

Since $(y_n^-)^i - (x_n^-)^i \leq (y_n^-)^i$ for all i we have

$$\varepsilon \leq \|x_n^+ - y_n^+\|_p + \|y_n^-\|_p \leq \|x_n^+ - y_n^+\|_p + 1 \to 1.$$

Thus $\varepsilon_0(\ell^{p,\infty}) \leq 1$.

To see the other inequality, let $u = e_1 - e_2$ and $v = e_1$, then $\|(u + v)/2\|_{p,\infty} = 1$ and $\|u - v\|_{p,\infty} = 1$. Thus $\varepsilon_0(\ell^{p,\infty}) \geq 1$.

On the other hand considering the vectors e_1 and $-e_2$ it is clear that $\varepsilon_0(\ell^{p,1}) \geq 2^{1/p}$. In fact it can be proved that $\varepsilon_0(\ell^{p,1}) = 2^{1/p}$.

So spaces $\ell^{p,1}$ and $\ell^{p,\infty}$ are not UC.

Since the Clarkson modulus only depends on 2-dimensional subspaces, different generalizations of this modulus involving subspaces with dimension $k > 2$ have been introduced respectively by Milman [M] and by Sullivan [Su].

DEFINITION 2.6 (MILMAN). *Let X be a Banach space. The modulus of k-uniform convexity associated with the space X, $\Delta^k_X(\varepsilon)$, is defined as*

$$\Delta^k_X(\varepsilon) = \inf_{\|x\|=1} \inf_{\substack{V \subseteq X \\ \dim V = k}} \sup_{\substack{\|y\|=1 \\ y \in V}} (\|x + \varepsilon y\| - 1).$$

The space is said to be k-uniformly convex, k-UC for short, if $\Delta^k_X(\varepsilon) > 0$ for all $\varepsilon > 0$.

DEFINITION 2.7 (SULLIVAN). *Let X be a Banach space. The modulus of k-uniform rotundity associated with the space X, $\delta^k_X(\varepsilon)$, is defined as*

$$\delta^k_X(\varepsilon) = \inf\left\{ 1 - \left\| \sum_{i=1}^{k+1} \frac{x_i}{k+1} \right\| : \|x_i\| = 1 \, \forall i = 1, \ldots, k+1 \text{ and} \right.$$

$$\left. A(x_1, \ldots, x_{k+1}) \geq \varepsilon \right\},$$

where $A(x_1, \ldots, x_{k+1})$ is the k-dimensional volume of $co(x_1, \ldots, x_{k+1})$, that is,

$$A(x_1, \ldots, x_{k+1}) = \sup\left\{ \det \begin{vmatrix} 1 & \cdots & 1 \\ f_1(x_1) & \cdots & f_1(x_{k+1}) \\ \vdots & \vdots & \vdots \\ f_k(x_1) & \cdots & f_k(x_{k+1}) \end{vmatrix} \right.$$

$$\left. : f_i \in X^* \text{ and } \|f_i\| \leq 1 \forall i = 1, \ldots, k \right\}.$$

The space is said to be k-uniformly rotund, k-UR for short, if $\delta^k_X(\varepsilon) > 0$ for all $\varepsilon > 0$.

Notice that for the case $k = 1$ we have the spaces uniformly convex.

We begin by proving a theorem which shows that the notion of k-UR is "coherent".

THEOREM 2.8. *If a Banach space X is k-UR, then it is $(k+1)$-UR.*

Proof. Suppose that there are norm one sequences $\{x_n^{(1)}\}, \ldots, \{x_n^{(k+2)}\}$ with

$$\lim_{n \to \infty} \|x_n^{(1)} + \cdots + x_n^{(k+2)}\| = k + 2.$$

Then from the fact that $\|x_n^{(j)}\| = 1$ for all $j = 1, \ldots, k+2$, we have

$$\lim_{n \to \infty} \|x_n^{(1)} + \cdots + x_n^{(j-1)} + x_n^{(j+1)} + \cdots + x_n^{(k+2)}\| = k + 1.$$

If the space X is k-UR then

$$\lim_{n\to\infty} A(x_n^{(1)} + \cdots + x_n^{(j-1)} + x_n^{(j+1)} + \cdots + x_n^{(k+2)}) = 0, \tag{1}$$

for all $j = 1, \ldots, k+2$.

Now, let $f_1, \ldots, f_{k+1}$ be norm one functionals. Considering the expansion in minors along the second row of the determinant

$$\det \begin{vmatrix} 1 & \cdots & 1 \\ f_1(x_n^{(1)}) & \cdots & f_1(x_n^{(k+2)}) \\ \vdots & \vdots & \vdots \\ f_{k+1}(x_n^{(1)}) & \cdots & f_{k+1}(x_n^{(k+2)}) \end{vmatrix},$$

and keeping (1) in mind, we may conclude

$$\lim_{n\to\infty} A(x_n^{(1)} + \cdots + \cdots + x_n^{(k+2)}) = 0,$$

and hence that the space is $(k+1)$-UR. $\qquad\square$

The following example due to Maluta and Prus [MP] shows that there exist spaces which are 2-UR but they are not UC.

Example 8: The space $(\mathbb{R}^2, \|\cdot\|)$, where

$$\|(x,y)\| = \frac{1}{2}\left((4x^2 + 5y^2)^{\frac{1}{2}} + |y|\right),$$

is a strictly convex space. Let $\|\cdot\|^*$ denote the corresponding norm of its dual. If we consider $X = (\mathbb{R}^3, \|\cdot\|)$, where

$$\|(x,y,z)\| = \|(\|(x,y)\|, z)\|^*,$$

the dual space is $X^* = (\mathbb{R}^3, \|\cdot\|^*)$, where

$$\|(x,y,z)\|^* = \|(\|(x,y)\|^*, z)\|.$$

Let f be the element of X^* corresponding to the vector $(1,0,-1/2)$ and the norm one vectors of X, $u = (1,-1/2,0)$, $v = (1,1/2,0)$. Then $f(u) = f(v) = 1$ which using Theorem 1.3. shows that X is not strictly convex and so is not UC.

Now let us suppose that this space is not 2-UR. Then its unit sphere contains a triangle of positive area. Therefore $S(0,1)$ contains also a nontrivial segment with endpoints of the form $u_1 = (x_1, y_1, z)$, $u_2 = (x_2, y_2, z)$. We set $a_i = \|v_i\|$, where $v_i = (x_i, y_i)$ for $i = 1, 2$. It is easy to see that $v_1 \neq \lambda v_2$ for every positive scalar and hence $\|v_1 + v_2\| < a_1 + a_2$. So

$$1 = \left\|\frac{u_1 + u_2}{2}\right\| < \|(\frac{a_1 + a_2}{2}, z)\|^* \leq \frac{1}{2}(\|u_1\| + \|u_2\|) = 1.$$

Thus the space X is 2-UR.

The spaces $\ell^{p,1}$ are 2-UR [DLX2]. Since in example 7 we have proved that these spaces are not UC, $\ell^{p,1}$ also provide with examples of 2-UR spaces which are not UC.

Now we are going to obtain some lower bounds for the volume. In particular we will show that the volume is greater than or equal to the "height" multiplied by the "area" of the base.

LEMMA 2.9. *If* $x_1, \ldots, x_k \in X$, *then*

$$A(x_1, \ldots, x_{k+1}) \geq d(x_1, [x_2, \ldots, x_{k+1}]) A(x_2, \ldots, x_{k+1}),$$

where $[x_2, \ldots, x_{k+1}]$ *denotes the affine hull of* $\{x_2, \ldots, x_{k+1}\}$, *that is,*

$$[x_2, \ldots, x_{k+1}] = \left\{ y \in X : y = \sum_{i=2}^{k+1} \alpha_i x_i \ and \ \sum_{i=2}^{k+1} \alpha_i = 1 \right\}.$$

Proof. Let $x_1, \ldots, x_{k+1}$ be in X. From the definition we have

$$A(x_1, \ldots, x_{k+1}) = \sup \left\{ \det \begin{vmatrix} 1 & \cdots & 1 \\ f_1(x_1) & \cdots & f_1(x_{k+1}) \\ \vdots & \vdots & \vdots \\ f_k(x_1) & \cdots & f_k(x_{k+1}) \end{vmatrix} \right.$$

$$\left. : f_i \in X^* \text{and } \|f_i\| \leq 1 \text{ for } i = 1, \ldots, k \right\}$$

$$= \sup \{ M_1 f_k(x_1) + \cdots + M_{k+1} f_k(x_{k+1}) : f_i \in X^*$$
$$\text{and } \|f_i\| \leq 1 \text{ for } i = 1, \ldots, k \}$$

$$= \sup \left\{ f_k \left(\sum_{i=1}^{k+1} M_i x_i \right) : f_i \in X^* \text{and } \|f_i\| \leq 1 \text{ for } i = 1, \ldots, k \right\}$$

$$= \sup \left\{ \left\| \sum_{i=1}^{k+1} M_i x_i \right\| : f_i \in X^* \text{and } \|f_i\| \leq 1 \text{ for } i = 1, \ldots, k-1 \right\}$$

where M_i is the minor corresponding to the element $f_k(x_i)$. Notice that $M_1 + \cdots + M_{k+1} = 0$ and $f_1, \ldots, f_{k-1}$ can be chosen so that M_1 is close to $A(x_2, \ldots, x_{k+1})$ and hence we obtain

$$A(x_1, \ldots, x_{k+1}) \geq \|x_1 + \alpha_2 x_2 + \cdots + \alpha_{k+1} x_{k+1}\| A(x_2, \ldots, x_{k+1})$$

where $\sum_{i=2}^{k+1} \alpha_i = -1$. Therefore

$$A(x_1, \ldots, x_{k+1}) \geq d(x_1, [x_2, \ldots, x_{k+1}]) A(x_2, \ldots, x_{k+1}).$$

$\square$

COROLLARY 2.10. *For all vectors* $x_1, \ldots, x_{k+1}$ *in* X *we have*

$$A(x_1, \ldots, x_{k+1}) \geq \prod_{i=1}^{k} d_i,$$

where

$$d_i = d(x_i, [x_{i+1}, \ldots, x_{k+1}]).$$

Geremia and Sullivan in [GeS] have proved that the definition of 2-uniform rotundity and 2-uniform convexity are equivalent. The general case was studied by [Ln2].

We will start with some geometric lemmas. The first is of a general character and was proved by Phelps [Ph].

LEMMA 2.11. *Suppose that* X *is a normed linear space and that* $\varepsilon > 0$. *If* f, g *are norm one functionals such that* $f^{-1}(0) \cap \overline{B}(0,1) \subset g^{-1}([-\varepsilon/2, \varepsilon/2])$, *then either* $\|f - g\| \leq \varepsilon$ *or* $\|f + g\| \leq \varepsilon$.

Proof. From the Hahn-Banach theorem we can choose $h \in X^*$ such that $h = g$ on $f^{-1}(0)$ and $\|h\| = \sup\{\|g(x)\| : x \in f^{-1}(0) \cap \overline{B}(0,1)\} \leq \varepsilon/2$. Since $\ker f \subset \ker (g - h)$, there exists a real number α such that $g - h = \alpha f$. Hence $\|g - \alpha f\| \leq \varepsilon/2$.

We will assume that α is positive and will show that in this case $\|f - g\| \leq \varepsilon$. If α were a negative number we would obtain $\|f + g\| < \varepsilon$.

Case $\alpha \geq 1$:

$\alpha = \|\alpha f\| \leq \|g\| + \|g - \alpha f\|$ hence $1 - \alpha^{-1} \leq (1 + \|g - \alpha f\|)^{-1}\|g - \alpha f\| \leq \|g - \alpha f\|$.

So

$$\|g - f\| = \|(1 - \alpha^{-1})g + \alpha^{-1}(g - \alpha f)\|$$
$$\leq 1 - \alpha^{-1} + \alpha^{-1}\|g - \alpha f\| \leq 2\|g - \alpha f\| \leq \varepsilon.$$

Case $0 \leq \alpha < 1$:

$$\|g - f\| = \|g - \alpha f\| + \|(1 - \alpha)f\| = \|g - \alpha f\| + (1 - \alpha)$$
$$= \|g - \alpha f\| + \|g\| - \|\alpha f\| \leq 2\|g - \alpha f\| \leq \varepsilon.$$

$\square$

LEMMA 2.12. *If* $\dim(X) \geq 2$, *then for each* $\varepsilon > 0$ *there exist* $v_1, v_2, v_3 \in X$ *such that* $\|v_i\| = \varepsilon$, $i = 1, 2, 3$, $\sum_{i=1}^{3} v_i = 0$ *and* $d(v_1, \, \text{span}\{v_2\}) \geq \varepsilon/3$.

Proof. Let v_1 be a arbitrary vector of norm ε and consider the continuous function on the ε-sphere defined by

$$F(x) = \frac{\|v_1 + x\|}{3}.$$

Since $F(v_1) = 2\varepsilon/3$ and $F(-v_1) = 0$ there exists a vector v_2 of norm ε such that $F(v_2) = \varepsilon/3$. If we consider $v_3 = -v_1 - v_2$ then $\|v_i\| = \varepsilon$, $i = 1, 2, 3$, $\sum_{i=1}^{3} v_i = 0$ and so we only have to show that $d(v_1, \text{span}\{v_2\}) \geq \varepsilon/3$.

It would be enough to prove that there exists a norm one functional such that $f(v_2) = 0$ and $f(v_1) \geq \varepsilon/3$. Indeed, then $\|v_1 - \alpha v_2\| \geq f(v_1 - \alpha v_2) \geq \varepsilon/3$ for any $\alpha \in \mathbb{R}$. Assume $f(v_1) < \varepsilon/3$ for every $f \in \overline{B}(0, 1)$ such that $f(v_2) = 0$. Then from Lemma 2.11 it follows that $\|v_1 + v_2\| < 2\varepsilon/3$ or $\|v_1 - v_2\| < 2\varepsilon/3$. However $\|v_1 + v_2\| = \| - v_3\| = \varepsilon$ and $\|v_1 - v_2\| = \|2v_1 - v_3\| \geq \varepsilon$. So there exists a norm one functional of the required form and the proof is concluded. $\qquad\square$

LEMMA 2.13. *If $dim(X) = k$, then for each $\varepsilon > 0$ there exist $v_1, v_2, \ldots, v_{k+1} \in X$ such that $\|v_i\| = \varepsilon$, $i = 1, \ldots, k + 1$, $\sum_{i=1}^{k+1} v_i = 0$ and $d(v_i, [v_1, \ldots, v_{i-1}]) \geq \varepsilon/3$ for $2 \leq i \leq k$.*

Proof. There exist k norm one vectors in X, $\{x_i\}_{i=1}^{k}$ and k norm one vectors in X^*, $\{x_i^*\}_{i=1}^{k}$ such that $x_i^*(x_j) = \delta_i^j$. We use induction and the previous lemma to prove that there exist k vectors in X which satisfy the following properties:

(i) $v_{i+1} \in \text{span}\left(\left\{\sum_{j=1}^{i} v_j, x_{i+1}\right\}\right)$ $\qquad\qquad\qquad 1 \leq i \leq k - 1,$

(ii) $\|v_{i+1}\| = \varepsilon = \left\|\sum_{j=1}^{i} v_j\right\| = \left\|\sum_{j=1}^{i+1} v_j\right\|$ $\qquad\qquad 1 \leq i \leq k - 1,$

(iii) $\varepsilon/3 \leq d\left(v_{i+1}, \text{span}\left(\left\{\sum_{j=1}^{i} v_j\right\}\right)\right)$ $\qquad\qquad 1 \leq i \leq k - 1.$

Let $v_1 = \varepsilon x_1$ and suppose that $v_1, \ldots, v_i$, $i < k$, have been chosen such that they satisfy (i), (ii) and (iii). Since $\varepsilon = \left\|\sum_{j=1}^{i} v_j\right\|$, by the proof of Lemma 2.12 there exists $v_{i+1} \in \text{span}\left(\left\{\sum_{j=1}^{i} v_j, x_{i+1}\right\}\right)$ such that $\|v_{i+1}\| = \varepsilon = \left\|\sum_{j=1}^{i} v_j\right\| = \left\|\sum_{j=1}^{i+1} v_j\right\|$ and $\varepsilon/3 \leq d\left(v_{i+1}, \text{span}\left(\left\{\sum_{j=1}^{i} v_j\right\}\right)\right)$.

Let $\{v_1, \ldots, v_k\}$ satisfy (i), (ii) and (iii). If we consider $v_{k+1} = -\sum_{j=1}^{k} v_j$ then

$\|v_{k+1}\| = \varepsilon$, $\sum_{j=1}^{k+1} v_j = 0$ and so it only remains to prove $d(v_i, [v_1, \ldots, v_{i-1}]) \geq \varepsilon/3$ for $2 \leq i \leq k$.

For $1 \leq i \leq k-1$ we have $v_{i+1} = a_{i+1} \left(\sum_{j=1}^{i} v_j \right) + b_{i+1} x_{i+1}$ and by (iii)

$\left\| v_{i+1} - a_{i+1} \left(\sum_{j=1}^{i} v_j \right) \right\| \geq \varepsilon/3$ therefore $|b_{i+1}| \geq \varepsilon/3$. Notice that

$$\left\| v_{i+1} - \sum_{j=1}^{i} \alpha_j v_j \right\| = \left\| a_{i+1} \left(\sum_{j=1}^{i} v_j \right) + b_{i+1} x_{i+1} - \sum_{j=1}^{i} \alpha_j v_j \right\|$$

$$\geq \left| x_{i+1}^* \left(\sum_{j=1}^{i} (a_{i+1} - \alpha_j) v_j + b_{i+1} x_{i+1} \right) \right| = |b_{i+1}|$$

because $v_j \in \text{span}\{x_1, \ldots, x_j\}$. So $d(v_{i+1}, \text{span}\{v_1, \ldots, v_i\}) \geq |b_{i+1}| \geq \varepsilon/3$. $\square$

THEOREM 2.14. *Let X be a Banach space. Then X is k-uniformly rotund if and only if X is k-uniformly convex.*

The theorem is a direct consequence of the two following lemmas.

LEMMA 2.15. *For all $\varepsilon > 0$*

$$\delta_X^k(\eta) \leq \frac{\Delta_X^k(\varepsilon)}{1 + \Delta_X^k(\varepsilon)}$$

where

$$\eta = \frac{(k+1)\varepsilon^k}{3^k (1 + \Delta_X^k(\varepsilon))^k}.$$

Proof. Since for each $\varepsilon > 0$

$$\Delta_X^k(\varepsilon) = \inf\{\Delta_E^k(\varepsilon) : E \subset X, \ \dim(E) = k+1\},$$

$$\delta_X^k(\varepsilon) = \inf\{\delta_E^k(\varepsilon) : E \subset X, \ \dim(E) = k+1\},$$

without any loss of generality we may assume X is a finite dimensional space. By definition there exists a norm one vector $u \in X$ and a k-dimensional subspace Y of X such that

$$\Delta_X^k(\varepsilon) = \sup_{\substack{\|y\|=1 \\ y \in Y}} (\|u + \varepsilon y\| - 1).$$

By the previous lemma there exist $k+1$ vectors $v_1, \ldots, v_{k+1}$ in Y such that $\|v_i\| = \varepsilon$, $\sum_{i=1}^{k+1} v_i = 0$ and $d(v_i, [v_1, \ldots, v_{i-1}]) \geq \varepsilon/3$ for $2 \leq i \leq k$.

Let
$$a = \frac{1}{1 + \Delta_X^k(\varepsilon)}$$

and $x_i = a(u + v_i)$, $i = 1, \ldots, k+1$. Then $\|x_i\| \leq 1$, $1 \leq i \leq k+1$ and

$$
\begin{aligned}
A(x_1, \ldots, x_{k+1}) &= a^k A(u + v_1, \ldots, u + v_{k+1}) \\
&= a^k A(v_1, \ldots, v_{k+1}) = a^k(k+1) A(v_1, \ldots, v_k, 0) \\
&\geq a^k(k+1)\|v_1\| \prod_{i=1}^{k-1} d(v_{i+1}, [v_1, \ldots, v_i]) \geq a^k(k+1)\left(\frac{\varepsilon}{3}\right)^k.
\end{aligned}
$$

Thus if
$$\eta = \frac{(k+1)\varepsilon^k}{3^k(1 + \Delta_X^k(\varepsilon))^k},$$

by definition

$$\delta_X^k(\eta) \leq 1 - \frac{\left\|\sum_{j=1}^{k+1} x_j\right\|}{k+1} = 1 - \frac{\|(k+1)au\|}{k+1} = 1 - \frac{1}{1 + \Delta_X^k(\varepsilon)}.$$

So
$$\delta_X^k(\eta) \leq \frac{\Delta_X^k(\varepsilon)}{1 + \Delta_X^k(\varepsilon)}$$

and the proof is concluded. $\square$

LEMMA 2.16. *For all $\varepsilon > 0$*

$$\Delta_X^k(\eta) \leq \frac{\delta_X^k(\varepsilon)}{1 - \delta_X^k(\varepsilon)}$$

where
$$\eta = \frac{\varepsilon}{(k+1)^{k+1}(1 - \delta_X^k(\varepsilon))}.$$

Proof. For each ε there exist vectors $x_1, \ldots, x_{k+1}$ and norm one functionals $f_1, \ldots, f_k$ such that

$$
\varepsilon = \det \begin{vmatrix} 1 & \cdots & 1 \\ f_1(x_1) & \cdots & f_1(x_{k+1}) \\ \vdots & \vdots & \vdots \\ f_k(x_1) & \cdots & f_k(x_{k+1}) \end{vmatrix}
$$

and

$$\left\| \sum_{i=1}^{k+1} x_i \right\| = (k+1)(1 - \delta_X^k(\varepsilon)).$$

Let $v_1, \ldots, v_k$ be in X, where we obtain the vector v_i by changing the $i+1$ row to $x_1, \ldots, x_{k+1}$ in the previous determinant and dividing by $(k+1)^k \| \sum_{i=1}^{k+1} x_i \|$. That is,

$$v_i = \left((k+1)^k \left\| \sum_{i=1}^{k+1} x_i \right\| \right)^{-1} \det \begin{vmatrix} 1 & \cdots & 1 \\ f_1(x_1) & \cdots & f_1(x_{k+1}) \\ \vdots & \vdots & \vdots \\ x_1 & \cdots & x_{k+1} \\ \vdots & \vdots & \vdots \\ f_k(x_1) & \cdots & f_k(x_{k+1}) \end{vmatrix}.$$

Consequently

$$f_j(v_i) = \frac{\varepsilon}{(k+1)^k \| \sum_{i=1}^{k+1} x_i \|}, \quad \text{if } i = j,$$

$$f_j(v_i) = 0 \qquad \qquad \text{if } i \neq j$$

and so

$$\|v_i\| \leq f_i(v_i) = \eta.$$

Let

$$u = \frac{\sum_{i=1}^{k+1} x_i}{\left\| \sum_{i=1}^{k+1} x_i \right\|}.$$

The vectors $v_1, \ldots, v_k$ are linearly independent, so we only need to prove that if $\| \sum_{i=1}^{k} a_i v_i \| = \eta$, then

$$\left\| u + \sum_{i=1}^{k} a_i v_i \right\| \leq \frac{1}{1 - \delta_X^k(\varepsilon)}.$$

Suppose $\| \sum_{i=1}^{k} a_i v_i \| = \eta$. For all j, $f_j(\sum_{i=1}^{k} a_i v_i) \leq \eta$. So $|a_i| \leq 1$. If $c_1, \ldots, c_{k+1}$ are the solutions of the linear system

$$y_1 + \cdots + y_{k+1} = k+1,$$

$$\sum_{i=1}^{k+1} f_j(x_i)(y_i - 1) = a_j \eta (k+1)(1 - \delta_X^k(\varepsilon)), \quad 1 \leq j \leq k$$

then

$$c_i = \frac{\eta(1 - \delta_X^k(\varepsilon))(k+1)}{A(x_1, \ldots, x_{k+1})} \det \begin{vmatrix} 1 & \cdots & 0 & 1 & \cdots & 1 \\ f_1(x_1) & \cdots & a_1 & f_1(x_{i+1}) & \cdots & f_1(x_{k+1}) \\ \vdots & \vdots & \vdots & \vdots & \vdots & \vdots \\ f_k(x_1) & \cdots & a_k & f_k(x_{i+1}) & \cdots & f_k(x_{k+1}) \end{vmatrix} + 1$$

$$\geq 1 - \frac{1}{(k+1)^k}(k+1)^{\frac{k+1}{2}} > 0.$$

(To obtain the last inequality we use *Hadamard's inequality*: If $|a_{i,j}| \leq 1$, then $|det(a_{i,j})_{k \times k}| \leq (k+1)^{\frac{k+1}{2}}$).

If we suppose

$$u + \sum_{i=1}^{k+1} a_i v_i = \frac{\sum_{i=1}^{k+1} c_i x_i}{\| \sum_{i=1}^{k+1} x_i \|}, \tag{2}$$

then

$$\left\| u + \sum_{i=1}^{k+1} a_i v_i \right\| = \frac{\| \sum_{i=1}^{k+1} c_i x_i \|}{\| \sum_{i=1}^{k+1} x_i \|}$$

$$\leq \frac{|\sum_{i=1}^{k+1} c_i|}{\| \sum_{i=1}^{k+1} x_i \|} \leq \frac{k+1}{\| \sum_{i=1}^{k+1} x_i \|} = \frac{1}{1 - \delta_X^k(\varepsilon)},$$

and the result would follow.

In order to prove (2) it is enough to find $(k+1)$ linearly independent functionals which take the same value on the two vectors.

If $x_1, \ldots, x_{k+1}$ are linearly independent then there exists f_0 such that $f_0(x_i) = 1$ for all $1 \leq i \leq k+1$. So $f_0(v_i) = 0$ for all $i = 1, \ldots, k$ and $f_0, \ldots, f_k$ are independent. Thus, it will be enough to prove for all $j = 0, \ldots, k$ that

$$f_j \left(u + \sum_{i=1}^{k+1} a_i v_i \right) = f_j \left(\frac{\sum_{i=1}^{k+1} c_i x_i}{\| \sum_{i=1}^{k+1} x_i \|} \right).$$

But this equality is true since

$$f_0 \left(u + \sum_{i=1}^{k+1} a_i v_i \right) = f_0 \left(\frac{\sum_{i=1}^{k+1} x_i}{\| \sum_{i=1}^{k+1} x_i \|} \right) =$$

$$= \frac{k+1}{\| \sum_{i=1}^{k+1} x_i \|} = f_0 \left(\frac{\sum_{i=1}^{k+1} c_i x_i}{\| \sum_{i=1}^{k+1} x_i \|} \right).$$

While for $j = 1, \ldots, k$

$$f_j \left(u + \sum_{i=1}^{k} a_i v_i \right) = f_j \left(\frac{\sum_{i=1}^{k+1} x_i}{\| \sum_{i=1}^{k+1} x_i \|} \right) + a_j \eta$$

and

$$f_j\left(\frac{\sum_{i=1}^{k+1} c_i x_i}{\|\sum_{i=1}^{k+1} x_i\|}\right) = \frac{\sum_{i=1}^{k+1} c_i f_j(x_i)}{\|\sum_{i=1}^{k+1} x_i\|}$$

$$= \frac{\sum_{i=1}^{k+1} f_j(x_i)}{\|\sum_{i=1}^{k+1} x_i\|} + \frac{a_j \eta(k+1)(1 - \delta_X^k(\varepsilon))}{\|\sum_{i=1}^{k+1} x_i\|}$$

$$= \frac{\sum_{i=1}^{k+1} f_j(x_i)}{\|\sum_{i=1}^{k+1} x_i\|} + a_j \eta.$$

In the case that $x_1, \ldots, x_{k+1}$ are linearly dependent, we have $[x_1, \ldots, x_{k+1}] = [v_1, \ldots, v_k]$ and then the dual is isometric to $[f_1, \ldots, f_k]$. So the proof is now concluded. $\qquad\square$

We now proceed to examining the consequences of a space being k-uniformly convex.

DEFINITION 2.17. *The characteristic of k-convexity $\varepsilon_0^k(X)$ of a Banach space X is defined as follows*

$$\varepsilon_0^k(X) = \sup\{\varepsilon : \delta_X^k(\varepsilon) = 0\}.$$

The next theorem was proved by Kirk [Ki3] and is concerned with the relationship between k-uniform convexity and reflexivity.

We will need the following lemma.

LEMMA 2.18. *Let $\{x_n\}$ be a bounded sequence in X for which $\|x_m - x_n\| \geq \varepsilon > 0$ if $m \neq n$. Then given any $k \in \mathbb{N}$ and any $\rho \in (0, \varepsilon)$ there exists a subsequence $\{z_n\}$ of $\{x_n\}$ such that if $z_{n_1}, \ldots, z_{n_{k+1}}$ are distinct points of $\{z_n\}$, then*

$$A(z_{n_1}, \ldots, z_{n_{k+1}}) \geq \varepsilon \left(\frac{\rho}{2}\right)^{k-1}.$$

Proof. The case $k = 1$ is trivial.

Let $\rho \in (0, \varepsilon)$ and $\rho < \theta < \varepsilon$. As we may suppose that the proposition holds for all $n \leq k - 1$, there exists a subsequence, again denoted by $\{x_n\}$, such that

$$A(x_{n_1}, \ldots, x_{n_k}) \geq \varepsilon \left(\frac{\theta}{2}\right)^{k-2}$$

whenever $n_j \neq n_i$ if $i \neq j$.

By countably ordering all k-tuples $\{i_1, \ldots, i_k : i_j \in \mathbb{N}, j = 1, \ldots, k\}$ it is possible to obtain a subsequence $\{x_{\psi(n)}\}$ of $\{x_n\}$, by thinning and diagonalization, such that for all $\{i_1, \ldots, i_k : i_j \in \mathbb{N}, j = 1, \ldots, k\}$ the limits,

$$\lim_{n \to \infty} A(x_{i_1}, \ldots, x_{i_k}, x_{\psi(n)}) = a(x_{i_1}, \ldots, x_{i_k}),$$

exist.

Our first step will be to prove that

$$\varepsilon \left(\frac{\theta}{2}\right)^{k-1} \leq \inf\{a(x_{i_1}, \ldots, x_{i_k}) : i_j \in \mathbb{N}, \, i_j \neq i_h \text{ if } j \neq h\}.$$

If this were not the case there would exist k distinct points of $\{x_n\}$ such that

$$\varepsilon \left(\frac{\theta}{2}\right)^{k-1} > \lim A(x_{i_1}, \ldots, x_{i_k}, x_{\psi(n)}) \geq \varepsilon \lim \, d(x_{\psi(n)}, [x_{i_1}, \ldots, x_{i_k}]) \left(\frac{\theta}{2}\right)^{k-2}.$$

Thus the above inequalities implie for large enough n that

$$d(x_{\psi(n)}, [x_{i_1}, \ldots, x_{i_k}]) < \frac{\varepsilon}{2}.$$

For each n there exists a point $y_{\psi(n)}$ of $[x_{i_1}, \ldots, x_{i_k}]$ where the distance is attained, that is

$$\|x_{\psi(n)} - y_{\psi(n)}\| = \, d(x_{\psi(n)}, [x_{i_1}, \ldots, x_{i_k}]) < \frac{\varepsilon}{2}.$$

Then from the fact that $\{x_n\}$ is a bounded set and the affine space $[x_{i_1}, \ldots, x_{i_k}]$ is finite dimensional we may deduce that there exists a subsequence of $\{y_{\psi(n)}\}$ which is convergent. By considering the corresponding subsequence of $\{x_{\psi(n)}\}$ we may deduce the existence of two points in $\{x_n\}$ with separation less than ε which is a contradiction.

Now, if we consider $n_i = i$ for $i = 1, \ldots, k$ there exists n_0 in $\mathbb{N}$ such that

$$A(x_{n_1}, \ldots, x_{n_k}, x_{\psi(n_0)}) \geq \varepsilon \left(\frac{\rho}{2}\right)^{k-1}.$$

Let $x_{n_{k+1}} = x_{\psi(n_0)}$ and for $s \geq k+1$ suppose n_s has been defined so that

$$A(y_1, \ldots, y_k, y_{k+1}) \geq \varepsilon \left(\frac{\rho}{2}\right)^{k-1}$$

for any set $\{y_1, \ldots, y_k, y_{k+1}\}$ of $k+1$ distinct points of $\{x_{n_1}, \ldots, x_{n_s}\}$. It is possible to repeat the argument, so there exists n_1 in $\mathbb{N}$ such that

$$A(x_{i_1}, \ldots, x_{i_k}, x_{\psi(n_1)}) \geq \varepsilon \left(\frac{\rho}{2}\right)^{k-1}$$

for any set $\{x_{i_1}, \ldots, x_{i_k}\}$ of k distinct points of $\{x_{n_1}, \ldots, x_{n_s}\}$. Let $x_{n_{s+1}} = x_{\psi(n_1)}$. If we consider the subsequence $\{x_{n_s}\}$ the proof is concluded. $\square$

We also use the following characterization of nonreflexivity due to James. The proof may be found in [B, Theorem 6, page 51]

LEMMA 2.19. *For a Banach space X, the following conditions are equivalent:*
(a) *X is not reflexive.*
(b) *For every θ, $0 < \theta < 1$, there is a sequence of points $\{x_n\}$ in X, with $\|x_n\| = 1$ for all n, and a sequence of linear functionals $\{f_n\}$, with $\|f_n\| = 1$ for all n, such that*

$$f_n(x_k) = \theta \quad \text{if} \quad n \leq k$$
$$f_n(x_k) = 0 \quad \text{if} \quad n > k.$$

(c) *For every θ, $0 < \theta < 1$, there is a sequence of points $\{x_n\}$ in X, with $\|x_n\| = 1$ for all n, such that:*
For all $K \geq 1$ and all $k \leq K$, if $\alpha_1, \ldots, \alpha_k, \alpha_{k+1}, \ldots, \alpha_K$ are positive numbers satisfying $\sum_{i=1}^k \alpha_i = \sum_{i=k+1}^K \alpha_i = 1$, then:

$$\left\| \sum_{i=1}^k \alpha_i x_i - \sum_{i=k+1}^K \alpha_i x_i \right\| \geq \theta.$$

THEOREM 2.20. *If $\varepsilon_0^k < (\frac{1}{2})^{k-1}$ for some $k \in \mathbb{N}$, then X is reflexive.*

Proof. If we suppose that X is not reflexive then, by James's characterization of nonreflexivity (Lemma 2.19), for every θ, $0 < \theta < 1$, there is a sequence of points $\{x_n\}$ in X with $\|x_n\| = 1$ for all n, and a sequence of linear functionals $\{f_n\}$, with $\|f_n\| = 1$ for all n, such that if $i > j$ then

$$\|x_i - x_j\| \geq f_j(x_i - x_j) = \theta.$$

Also if $z \in \mathrm{co}(\{x_1, \ldots, x_n\})$, $\|z\| \geq f_1(z) = \theta$. From Lemma 2.18, given $\rho \in (0, \theta)$ and $k \in \mathbb{N}$, we know $\{x_n\}$ has a subsequence, and we may take to be $\{x_n\}$, such that

$$A(x_1, \ldots, x_{k+1}) \geq \theta \left(\frac{\rho}{2} \right)^{k-1}.$$

It follows that $\delta_X^k \left(\theta \, (\rho/2)^{k-1} \right) \leq 1 - \theta$. Since ρ and θ are arbitrarily near to 1, we have $\varepsilon_0^k \geq (1/2)^{k-1}$. So if $\varepsilon_0^k < (\frac{1}{2})^{k-1}$, then X is reflexive. $\qquad \square$

As corollary of this theorem we deduce the following results which were first proved by Goebel [Go1] and Bernal and Sullivan [BS2].

COROLLARY 2.21. *If $\varepsilon_0(X) < 1$, then X is reflexive.*

COROLLARY 2.22. *If the space X is k-uniformly convex for some $k \in \mathbb{N}$, then X is reflexive.*

Another property of k-uniformly convex spaces which we study concerns the strong convergence of weakly convergent sequences in the unit sphere. Letting X be a Banach space. We will say that the norm is a Kadec-Klee norm, (KK norm for short) if and only if sequences converge strongly on the unit sphere whenever they converge weakly. An equivalent formulation is the following:

DEFINITION 2.23. *The norm of a Banach space X is said to be a Kadec-Klee norm, KK for short, if, for all sequences $\{x_n\}$ in the unit ball, we have*

$$\left.\begin{array}{r} \{x_n\}\,is\ not\ norm\ Cauchy \\ x_n \rightharpoonup x \end{array}\right\} \Rightarrow \|x\| < 1.$$

This formulation suggests the following stronger notion due to Huff [Hu]:

DEFINITION 2.24. *The norm of a Banach space X is said to be uniformly Kadec-Klee (UKK) if for each $\varepsilon > 0$ there exists $\delta > 0$ such that if $\{x_n\}$ is a sequence in the unit ball, then*

$$\left.\begin{array}{r} \inf\{\|x_n - x_m\| : n \neq m\} > \varepsilon \\ x_n \rightharpoonup x \end{array}\right\} \Rightarrow \|x\| < 1 - \delta.$$

In chapter V we shall consider again this notion in connection with a measure of noncompactness. Now, we shall study the connection with k-uniform convexity.

The following theorem was first proved by Yu [Y1]. Here we include a proof due to Kirk [Ki3].

THEOREM 2.25. *If X is k-uniformly convex for certain k, then the norm of X is UKK.*

Proof. If we suppose X is k-uniformly convex for some k and that a sequence $\{x_n\}$ is under the condition of the previous definition, then by Lemma 2.18, given $\rho \in (0, \varepsilon)$ and $k \in \mathbb{N}$, $\{x_n\}$ has a subsequence, we may take to be $\{x_n\}$, such that

$$A(x_n, \ldots, x_{n+k}) \geq \varepsilon(\frac{\rho}{2})^{k-1}$$

for $k = 1, 2, \ldots$

Thus

$$\left(\frac{1}{k+1}\right)\left\|\sum_{i=0}^{k} x_{n+i}\right\| \leq 1 - \delta_X^k\left(\varepsilon\left(\frac{\rho}{2}\right)^{k-1}\right).$$

Since the sequence

$$\left\{\left(\frac{1}{k+1}\right)\sum_{i=0}^{k} x_{n+i}\right\}$$

also converges weakly to x, we have the desired conclusion. $\qquad\square$

3. k-uniform smoothness

Using Milman's definition of k-uniform convexity, E. Maluta and S. Prus [MP] have introduced a dual concept, k-uniform smoothness, and have studied the relation between these notions.

DEFINITION 3.1. *A Banach space X is said to be k-uniformly smooth (k-US for short) if for every $\varepsilon > 0$, there exists $\eta > 0$ such that for every norm one vectors $x \in X$ and for every t, $0 < t < \eta$, if V is a k-dimensional subspace of X then there exists a norm one vector $y \in V$ such that*

$$\frac{1}{2}(\|x + ty\| + \|x - ty\|) < 1 + \varepsilon t.$$

THEOREM 3.2. *If a Banach space X is k-US for some $k \in \mathbb{N}$, then X is reflexive.*

Proof. Suppose X is not a reflexive space. By James's characterization of nonreflexivity (Lemma 2.19), if we take $\theta < 1$ there exists a norm one sequence $\{x_n\}$ in X and a sequence of linearly independent functionals $\{f_n\}$ such that

$$f_n(x_m) = \theta \quad \text{if} \quad n \leq m,$$
$$f_n(x_m) = 0 \quad \text{if} \quad n > m.$$

Let $\{x_1, x_2, \ldots, x_n\}$ and $\{f_1, f_2, \ldots, f_n\}$ be as above with $n = 2k + 1$. If we define $y_i = x_{2i} - x_{2i-1}$, it follows that

$$f_{2i}(y_i) = \theta$$
$$f_{2j}(y_i) = 0 \text{ when } j \neq i.$$

Furthermore, the new collection of points $\{y_i\}$ satisfies $\theta \leq \|y_i\| \leq 2$.

Now, we can write $z_i = y_i/\|y_i\|$ and $z = x_{2k+1}$.

Consider $V = \text{span}(\{z_1, z_2, \ldots, z_k\})$ and $v \in V$ with norm equal to one. So $v = \sum_{i=1}^{k} \alpha_i z_i$. If we let j such that $|\alpha_j| = \max\{|\alpha_i| : i = 1, \ldots, k\}$, then $|\alpha_j| \geq 1/k$.

Now we only need to evaluate $\frac{1}{2}(\|z + tv\| + \|z - tv\|)$ as follows:

$$\frac{1}{2}(\|z + tv\| + \|z - tv\|) \geq \frac{1}{2}(f_{2j}(z + tv) + f_{2k+1}(z - tv))$$

and so if $\alpha_j > 0$,

$$\frac{1}{2}(\|z + tv\| + \|z - tv\|) \geq \frac{1}{2}(\theta + \frac{t\alpha_j \theta}{2} + \theta) = \theta + \frac{t\alpha_j \theta}{4} \geq \theta(1 + \frac{t}{4k}).$$

Choosing an appropriate θ we therefore have

$$\frac{1}{2}(\|z + tv\| + \|z - tv\|) \geq 1 + \frac{t}{8k}.$$

The case $\alpha_j < 0$ is similar, and from here, we readily see that X is not k-US. $\square$

THEOREM 3.3. *Let X be a Banach space. Then*
   (a) *X is k-uniformly convex if and only if X^* is k-uniformly smooth.*
   (b) *X is k-uniformly smooth if and only if X^* is k-uniformly convex.*

Proof. We will prove statement (b). Statement (a) follows from statement (b) and Theorem 3.2. In the first step of our proof we show that if X is not k-US then X^* is not k-UR.

If X is not k-US then there exists $\varepsilon > 0$ such that for every $\eta > 0$ there exists t, $0 < t < \eta$, a norm one vector $x \in X$ and a k-dimensional subspace V, such that for all norm one vectors y in V

$$\frac{1}{2}(\|x + ty\| + \|x - ty\|) \geq 1 + \varepsilon t. \tag{3}$$

Let x^* be a norm one vector in X^* such that $x^*(x) = 1$. We define $x_1^* = x^*$.

Notice that the vector x is not in V, as otherwise $y = -x$ would not verify (3). Hence $\dim(\operatorname{span}\{x \cup V\}) = k + 1$. Let $W_1 = \ker x^* \cap \operatorname{span}\{x \cup V\}$, then $\dim W_1 = k$.

If $z \in W_1$, $\|z\| = 1$, then $z = u - \lambda x$ for some $\lambda \in \mathbb{R}$ and $u \in V$. So $u = \lambda x + z \in V$ and

$$\left.\begin{array}{l} \|u\| \geq |1 - |\lambda|| \\ \|u\| \geq |x^*(u)| = |\lambda| \end{array}\right\} \Rightarrow \|u\| \geq \frac{1}{2}.$$

Let $y = u/\|u\|$. Then we may find norm one vectors u^*, v^* in X^* such that

$$u^*(x + ty) = \|x + ty\|,$$
$$v^*(x - ty) = \|x - ty\|.$$

Hence

$$u^*(x + ty) + v^*(x - ty) \geq 2 + 2\varepsilon t$$
$$u^*(x) + v^*(x) \geq 2 + 2\varepsilon t - 2t$$

and since $u^*(x) \leq 1$, $v^*(x) \leq 1$, we obtain

$$u^*(x) > 1 - 2t(1 - \varepsilon)$$

and

$$v^*(x) > 1 - 2t(1 - \varepsilon).$$

Therefore $|u^*(x) - v^*(x)| \leq 2t(1 - \varepsilon)$. Let $s = t/\|u\|$. Then considering η such that $t \leq \varepsilon/(1 - \varepsilon)$ and noting that $|\lambda|s \leq \|u\|s = t$ we have

$$\begin{aligned} \|x + sz\| + \|x - sz\| &= \|x + ty - \lambda sx\| + \|x - ty + \lambda sx\| \\ &\geq u^*(x + ty - \lambda sx) + v^*(x - ty + \lambda sx) \\ &\geq \|x + ty\| - \lambda su^*(x) + \|x - ty\| + \lambda sv^*(x) \\ &\geq 2 + 2\varepsilon t - \lambda s(u^*(x) - v^*(x)) \geq 2 + 2\varepsilon t - |\lambda|s2t(1 - \varepsilon) \\ &\geq 2 + 2\varepsilon t - 2t^2(1 - \varepsilon) \geq 2\left(1 + \frac{t\varepsilon}{2}\right). \end{aligned}$$

Consequently, if we consider norm one functionals φ and ψ such that

$$\|x + sz\| = \varphi(x + sz), \qquad \|x - sz\| = \psi(x - sz),$$

and we suppose $(\varphi - \psi)(z) \geq 0$ (in the other case consider $(\psi - \varphi)(z)$) we obtain

$$\begin{aligned}
\|x + sz\| + \|x - sz\| &= (\varphi + \psi)(x) + s(\varphi - \psi)(z) \\
&\leq (\varphi + \psi)(x) + 2t(\varphi - \psi)(z) \leq \|x + 2tz\| + \|x - 2tz\|.
\end{aligned}$$

So

$$\|x + 2tz\| + \|x - 2tz\| \geq 2\left(1 + \frac{t\varepsilon}{2}\right).$$

Let us now fix $z_1 \in W_1$, $\|z_1\| = 1$. Then either

$$\|x + 2tz_1\| \geq 1 + \frac{t\varepsilon}{2}$$

or

$$\|x - 2tz_1\| \geq 1 + \frac{t\varepsilon}{2}.$$

Assume that the first inequality holds. In this case we find a norm one functional $x_2^* \in X^*$ for which

$$x_2^*(x + 2tz_1) = \|x + 2tz_1\|.$$

Clearly $x_2^*(z_1) \geq \varepsilon/4$ and $x_2^*(x) \geq 1 - ct$, where $c = 2 - \varepsilon/2$. Notice that

$$\|\alpha x_1^* - x_2^*\| \geq |\alpha x_1^*(z_1) - x_2^*(z_1)| = |x_2^*(z_1)| \geq \varepsilon/4$$

for every $\alpha \in \mathbb{R}$. Therefore $d(x_2^*, [x_1^*]) \geq \varepsilon/4$.

Now consider the subspace $\ker x_2^* \cap W_1 = W_2$. Clearly $\dim W_2 = k - 1$. Repeating the above argument with z_1 replaced by a norm one vector $z_2 \in W_2$, we obtain a norm one functional $x_3^* \in X^*$ such that $x_3^*(z_2) \geq \varepsilon/4$ and $x_3^*(x) \geq 1 - ct$. From the first inequality it follows that $d(x_3^*, [x_1^*, x_2^*]) \geq \varepsilon/4$.

Proceeding in this way, we get norm one functionals $x_1^*, \ldots, x_{k+1}^*$ satisfying the conditions:

$$d(x_{j+1}^*, [x_1^*, \ldots, x_j^*]) \geq \frac{\varepsilon}{4}, \qquad j = 1, \ldots, k,$$

$$\frac{1}{k+1}\|x_1^* + \cdots + x_{k+1}^*\| \geq 1 - ct.$$

It only remains to prove that if X is not k-UC then X^* is not k-US.

If the space X is not k-uniformly convex then there exists $\varepsilon \in (0, 1)$ such that for every $\delta > 0$ there exists a norm one vector $x \in X$ and a subspace V, $\dim V = k$ such that for every norm one vector $y \in V$ we have $\|x + \varepsilon y\| < 1 + \delta$.

We take an arbitrary $t \in (0,1)$ and put $\delta = \varepsilon^2 t/18k$. The assumption gives us a corresponding element x and a k-dimensional subspace V. Let $y_1, \ldots, y_k$ be an *Auerbach system* of V, that is, n vectors $y_1, \ldots, y_k$ of norm one in V and n vectors $y_1^*, \ldots, y_k^*$ of norm one in V^* so that $x_j^*(x_i) = \delta_i^j$ (see [LT1, page 16]). We consider the vectors $u_i = x + \varepsilon y_i$, $i = 1, \ldots, k$. From the assumption the space X is not k-uniformly convex, we obtain

$$\|x \pm \varepsilon y_i\| < 1 + \delta.$$

The first step will be to prove that

$$d(u_i, \ \text{span}\{x, \{u_j\}_{j \neq i}\}) \geq a, \tag{4}$$

where $a = \varepsilon^2/3$.

For this purpose we take $y \in V$ with $\|y\| = 1$. Then

$$\| - sx + y\| \geq |s| - 1$$

and

$$\|y - sx\| = \|y + |s|\varepsilon y - sx - |s|\varepsilon y\|$$

$$\geq (1 + |s|\varepsilon) - |s|(1 + \delta) > 1 - |s|\left(1 - \frac{2}{3}\varepsilon\right)$$

for every $s \in \mathbb{R}$. This shows that

$$\|y - sx\| \geq \frac{\varepsilon}{3}.$$

Let us observe that the above inequality also holds in the case when $\|y\| \geq 1$.

Since $y_1, \ldots, y_n$ is an Auerbach system, we see that

$$\left\| \varepsilon y_i - \sum_{j \neq i} \varepsilon \alpha_j y_j \right\| \geq \varepsilon$$

for all scalars $\alpha_1, \ldots, \alpha_k$, and it follows that

$$\left\| u_i - \alpha_0 x + \sum_{j \neq i} \alpha_j u_j \right\| \geq a.$$

Thus we have established inequality (4).

Using (4), we can find norm one functionals $u_1^*, \ldots, u_k^*$ in X^* such that $u_i^*(u_i) \geq a$, $u_i^*(u_j) = 0$ if $j \neq i$ and $u_i^*(x) = 0$ for $i = 1, \ldots, k$. Consequently $u_i^*(y_j) = 0$ whenever $j \neq i$.

Consider now a norm one functional x^* such that $x^*(x) = 1$. We have

$$a \leq u_i^*(x + \varepsilon y_i) = \varepsilon u_i^*(y_i).$$

Consequently $u_i^*(y_i) \geq \frac{\varepsilon}{3}$ for $i = 1, \ldots, k$. Moreover

$$1 \pm \varepsilon x^*(y_i) = x^*(x \pm \varepsilon y_i) \leq \|x \pm \varepsilon y_i\| < 1 + \delta.$$

Therefore $|x^*(y_i)| < \dfrac{\delta}{\varepsilon}$ for $i = 1, \ldots, k$. We take $y^* = \sum_{i=1}^{k} \beta_i u_i^*$, $\|y^*\| = 1$. Then

$$1 \leq \sum_{i=1}^{k} |\beta_i| \leq k \max |\beta_i| = k|\beta_j|$$

for some index j.

From this we obtain

$$\frac{1}{2}(\|x^*+ty^*\|+\|x^*-ty^*\|) \geq \frac{1}{2}\left(\frac{1}{1+\delta}(x^*+ty^*)(x+(\operatorname{sgn}\beta_j)\varepsilon y_j)+(x^*-ty^*)(x)\right)$$

$$\geq \frac{1}{2}\left(\frac{1}{1+\delta}(1+ta|\beta_j|-\delta)+1\right) > 1 + \frac{a}{4k}t,$$

and so X^* is not k-US. $\qquad\qquad\qquad\qquad\qquad\qquad\qquad\qquad\qquad\qquad\qquad\qquad\square$

Remark 3.4. If $1 < p < \infty$ then ℓ^p is 1-UC and so 1-US.

COROLLARY 3.5. *If a Banach space X is k-US, then X is $(k+1)$-US.*

Remark 3.6. Bearing in mind Theorem 3.3 and Example 8, $\ell^{p,\infty}$ provides with an example of a space which is 2-US but is not 1-US.

DEFINITION 3.7. *Let X be a Banach space. We define the modulus of k-uniform smoothness of X by*

$$\beta_X^k(t) = \sup_{\|x\|=1} \sup_{\substack{V \subset X \\ \dim V = k}} \inf_{\substack{\|y\|=1 \\ y \in V}} \left\{\frac{1}{2}(\|x+ty\|+\|x-ty\|)-1\right\}.$$

DEFINITION 3.8. *A Banach space X is said to be k-uniformly smooth if and only if*

$$\lim_{t \to 0^+} \frac{\beta_X^k(t)}{t} = 0.$$

Remark 3.9. For the case $k = 1$, $\beta_X^k(t)$ is the classical *modulus of smoothness* usually denoted by ρ and defined

$$\rho_X(t) = \sup\left\{\frac{1}{2}(\|x+ty\|+\|x-ty\|)-1 : \|x\| \leq 1, \|y\| \leq t\right\}.$$

1-US spaces are usually called *uniformly smooth spaces*, US for short.

There are relationships between the Clarkson modulus of a space X and the smoothness modulus of its dual and viceversa which are quantitative versions of the previous theorem. The proof of this result can be found in [B, page 208].

THEOREM 3.10 (LINDENSTRAUSS' DUALITY FORMULA). *For every Banach space X and every $t \geq 0$*

(a)
$$\rho_{X^*}(t) = \sup_{0 \leq \varepsilon \leq 2} \left\{ \frac{t\varepsilon}{2} - \delta_X(\varepsilon) \right\}.$$

(b)
$$\rho_X(t) = \sup_{0 \leq \varepsilon \leq 2} \left\{ \frac{t\varepsilon}{2} - \delta_{X^*}(\varepsilon) \right\}.$$

Remark 3.11. Using the Lindenstrauss' duality formula and the value of δ_{ℓ^2} (Example 6), we may deduce

$$\rho_{\ell^2}(t) = \sqrt{1 + t^2} - 1.$$

If we define the characteristic of smoothness $\rho_0(X)$ of a Banach space X as

$$\rho_0(X) = \lim_{t \to 0^+} \frac{\rho_X(t)}{t},$$

then we have the following result:

COROLLARY 3.12. *For every Banach space X we have*
(a) $\rho_0(X^*) = \varepsilon_0(X)/2$.
(b) $\rho_0(X) = \varepsilon_0(X^*)/2$.

Proof. We will only prove statement (a); the proof of statement (b) follows the same argument using the dual formula.

First notice that

$$\rho_0(X^*) = \lim_{t \to 0^+} \frac{\rho_{X^*}(t)}{t} = \lim_{t \to 0^+} \frac{1}{t} \sup_{0 \leq \varepsilon \leq 2} \left\{ \frac{t\varepsilon}{2} - \delta_X(\varepsilon) \right\}$$

$$= \lim_{t \to 0^+} \sup_{0 \leq \varepsilon \leq 2} \left\{ \frac{\varepsilon}{2} - \frac{\delta_X(\varepsilon)}{t} \right\} \geq \varepsilon_0(X)/2.$$

On the other hand, let

$$a = \lim_{t \to 0^+} \frac{\rho_{X^*}(t)}{t}.$$

Given $\eta > 0$ there exists $t_0 > 0$ such that for all t, $0 < t < t_0$, there exists $\varepsilon(t)$ satisfying the inequality

$$a - \eta < \frac{\varepsilon(t)}{2} - \frac{\delta_X(\varepsilon(t))}{t}.$$

From this inequality we can deduce $\varepsilon(t) > 2(a - \eta)$.

If we suppose $\delta_X(2(a - \eta)) > 0$ then $\delta_X(\varepsilon(t)) \geq \delta_X(2(a - \eta))$ and taking limit as $t \to 0^+$, we obtain the contradiction $a = -\infty$. Hence $\delta_X(2(a - \eta)) = 0$ and since η is arbitrary we obtain $\varepsilon_0(X)/2 \geq a$. $\square$

COROLLARY 3.13. *Let X be a Banach space with $\rho_0(X) < 1/2$. Then X is reflexive.*

Chapter V
Nearly Uniform Convexity and
Nearly Uniform Smoothness

Reflexivity and the uniform Kadec-Klee property are among the most important properties of k-uniformly convex spaces. The study of spaces satisfying both properties was initiated by Huff in 1980 [Hu] who called these spaces nearly uniformly convex.

A generalization of the moduli of k-uniform convexity, considering infinite dimensional subspaces, was given by Goebel and Sękowski in 1984 [GS]. The corresponding concept to k-uniformly convex space is, in this case, that of nearly uniformly convex space. Goebel and Sękowski use the Kuratowski measure of noncompactness as a measure of the volume of the sets. The use of the measure of noncompactness is relevant since it implies a strong connection between topological methods and metric methods.

We are going to start this chapter with the definition of a nearly uniformly convex Banach space and the study of the moduli which are defined using the measures of noncompactness.

In the second section of the chapter we will pay attention to nearly uniformly smooth spaces. These spaces have been introduced and studied by Prus [Pr5] who proved the duality between near uniform smoothness and near uniform convexity.

The last section is devoted to the study of the uniform Opial condition.

Zdzislaw Opial (1930–1974) was a Polish mathematician. He studied mathematics at the Jagiellonian University in Cracow from 1949 till 1954. His interests were wide-ranging and not restricted to mathematics; he was interested in arts, films and the theatre. As a student he played football in one of the local league teams.

In the second year of his university studies, Opial attended Wazewski's seminar on differential equations. This seminar was crucial in forming Opial's research interest.

During all his professional life Opial was associated with the Jagiellonian University. In 1951, while still a student, he was appointed to the position of assistant. He received a doctor's degree in 1957. He was promoted to the position of docent in 1962 and became a professor in

1967. He spent the year 1959–60 in Paris supported by the Polish Academy of Sciences. During the years 1966–67 he was a visiting professor at the Lefschetz Center for Dynamical Systems of Brown University in Providence, Rhode Island.

The Opial property was introduced in [O1, 1967] and has proved to be very useful in fixed point theory for nonexpansive mappings.

1. Nearly Uniformly Convex Banach Spaces

DEFINITION 1.1. *A Banach space X is said to be nearly uniformly convex (NUC) if for every $\varepsilon > 0$ there exists $\delta > 0$ such that if $\{x_n\}$ is a sequence in X, then*

$$\left. \begin{array}{c} \inf\{\|x_n - x_m\| : n \neq m\} > \varepsilon \\ \|x_n\| \leq 1 \end{array} \right\} \Rightarrow co(\{x_n\}) \cap B(0, 1 - \delta) \neq \emptyset.$$

Remark 1.2. From the definition it is clear that every UC space is NUC. Furthermore from Eberlein's theorem, it follows that a Banach space X is NUC if X is reflexive and UKK. In fact, Huff [Hu] proved that NUC is equivalent to UKK and reflexivity. We shall include a proof here using the moduli of noncompact convexity which we are going to introduce.

DEFINITION 1.3. *Let X be a Banach space and ϕ a measure of noncompactness on X. We define the modulus of noncompact convexity associated to ϕ in the following way*

$$\begin{aligned} \Delta_{X,\phi}(\varepsilon) &= \inf\{1 - d(0, A) : A \subset \overline{B}(0, 1) \text{ is convex }, \ \phi(A) > \varepsilon\} \\ &= \inf\{1 - d(0, \ co(A)) : A \subset \overline{B}(0, 1), \phi(A) > \varepsilon\}. \end{aligned}$$

We define the NUC-characteristic of X associated to the measure of noncompactness ϕ to be

$$\varepsilon_\phi(X) = \sup\{\varepsilon \geq 0 : \Delta_{X,\phi}(\varepsilon) = 0\}.$$

The function $\Delta_{X,\alpha}(\varepsilon)$ has been considered by Goebel and Sękowski [GS], $\Delta_{X,\chi}(\varepsilon)$ by Banaś [Ba1] and $\Delta_{X,\beta}(\varepsilon)$ by Domínguez and López [DL2].

Remark 1.4. The following relationships among the different moduli are easy to obtain

$$\delta_X(\varepsilon) \leq \Delta_{X,\alpha}(\varepsilon) \leq \Delta_{X,\beta}(\varepsilon) \leq \Delta_{X,\chi}(\varepsilon) \tag{1}$$

and consequently

$$\varepsilon_0(X) \geq \varepsilon_\alpha(X) \geq \varepsilon_\beta(X) \geq \varepsilon_\chi(X). \tag{2}$$

Example 1: Let $X = \ell^{2,1}$ the space introduced in Example II.3. We are going to prove that

$$\Delta_{X,\alpha}(\varepsilon) = 1 - \left(1 - \left(\frac{\varepsilon}{2}\right)^2\right)^{\frac{1}{2}}.$$

In Example II.3 it has been proved that for this space $\chi \equiv \mu$. Hence if a subset A of $\ell^{2,1}$ satisfies $\alpha(A) \geq \varepsilon$, then

$$\chi(A) = \lim_{n \to \infty} \|R_n A\| \geq \frac{\varepsilon}{2}.$$

Now, let us fix arbitrarily $\varepsilon \in [0,2)$ and A be a closed, convex set, $A \subset \overline{B}(0,1)$, with $\alpha(A) > \varepsilon$. Then A must contain a weakly convergent sequence, $x_n \rightharpoonup z$, for which

$$\liminf_{n \to \infty} \|R_n x_n\| \geq \frac{\varepsilon}{2}.$$

Let $\eta > 0$. Then for k sufficiently large $\|R_k z\| < \eta$ and for $n \geq k$

$$1 \geq \|x_n\|_{2,1} = \|x_n^+\|_2 + \|x_n^-\|_2$$
$$= \left(\|P_k x_n^+\|_2^2 + \|R_k x_n^+\|_2^2\right)^{\frac{1}{2}} + \left(\|P_k x_n^-\|_2^2 + \|R_k x_n^-\|_2^2\right)^{\frac{1}{2}},$$

and using the inequality

$$\sqrt{a^2 + b^2} + \sqrt{c^2 + d^2} \geq \sqrt{(a+c)^2 + (b+d)^2}$$

we obtain

$$1 \geq \left(\|P_k x_n^+\|_2 + \|P_k x_n^-\|_2\right)^2 + \left(\|R_k x_n^+\|_2 + \|R_k x_n^-\|_2\right)^2$$
$$\geq \|P_k x_n\|_{2,1}^2 + \|R_k x_n\|_{2,1}^2.$$

Letting $n \to \infty$ we have

$$1 \geq \|P_k z\|_{2,1}^2 + \left(\frac{\varepsilon}{2}\right)^2 \geq \|z\|_{2,1}^2 - \eta^2 + \left(\frac{\varepsilon}{2}\right)^2.$$

Thus

$$\Delta_{X,\alpha}(\varepsilon) \geq 1 - \left(1 - \left(\frac{\varepsilon}{2}\right)^2\right)^{\frac{1}{2}}.$$

To obtain the second inequality we consider the set

$$A = \left\{ x = \sum_{i=1}^{\infty} \lambda_i e_i : \|x\|_{2,1} \leq 1 \text{ and } \lambda_1 \geq \left(1 - \left(\frac{\eta}{2}\right)^2\right)^{\frac{1}{2}} \right\}$$

where $\eta > \varepsilon$ is arbitrary. The set A is closed, convex, $A \subset \overline{B}(0,1)$, $\alpha(A) \geq \eta > \varepsilon$ and

$$d(0,A) \geq \left(1 - \left(\frac{\eta}{2}\right)^2\right)^{\frac{1}{2}}.$$

So

$$\Delta_{X,\alpha}(\varepsilon) \leq 1 - \left(1 - \left(\frac{\eta}{2}\right)^2\right)^{\frac{1}{2}}.$$

Thus

$$\Delta_{X,\alpha}(\varepsilon) \leq 1 - \left(1 - \left(\frac{\varepsilon}{2}\right)^2\right)^{\frac{1}{2}}.$$

We are going to study the relationship between the value of $\varepsilon_\chi(X)$ and the reflexivity of X. Certain technical lemmas will be needed.

LEMMA 1.5. *If a Banach space X is nonreflexive, then there exists a decreasing sequence of closed, convex subsets $\{C_n\}$ in the unit ball $\overline{B}(0,1)$ such that*

$$\bigcap_{n=1}^{\infty} C_n = \emptyset.$$

Proof. Assume that X is nonreflexive. By James's characterization of the nonreflexivity (Lemma IV.2.19) we have, for all $0 < \theta < 1$, norm one sequences $\{x_n\}$ in X and norm one sequences $\{f_n\}$ in X^* such that

$$f_j(x_i) = \begin{cases} \theta & \text{if } i \geq j \\ 0 & \text{if } i < j. \end{cases}$$

Let $C_n = \overline{\text{co}}\{x_n\}_{k \geq n}$ and $x \in \bigcap_{n=1}^{\infty} C_n$. Since $x \in C_1$ if $\varepsilon < \theta/3$ there exist $N \in \mathbb{N}$ and $\lambda_1, \ldots, \lambda_N$ such that $\sum_{i=1}^{N} \lambda_i = 1$, $\lambda_i \geq 0$ and $\left\| x - \sum_{i=1}^{N} \lambda_i x_i \right\| < \varepsilon$. Since $x \in C_{N+1}$ there exist $M \in \mathbb{N}$ and $\eta_{N+1}, \ldots, \eta_{N+M}$ such that $\sum_{i=1}^{M} \eta_{N+i} = 1$, $\eta_{N+i} \geq 0$ and $\left\| x - \sum_{i=1}^{M} \eta_{N+i} x_{N+i} \right\| < \varepsilon$. Therefore

$$\theta = f_{N+1}\left(\sum_{i=1}^{M} \eta_{N+i} x_{N+i}\right) = f_{N+1}\left(\sum_{i=1}^{M} \eta_{N+i} x_{N+i} - \sum_{i=1}^{N} \lambda_i x_i\right) \leq$$

$$\leq \left\|\sum_{i=1}^{M} \eta_{N+i} x_{N+i} - \sum_{i=1}^{N} \lambda_i x_i\right\| \leq \left\| x - \sum_{i=1}^{M} \eta_{N+i} x_{N+i} \right\| + \left\| x - \sum_{i=1}^{N} \lambda_i x_i \right\| < 2\varepsilon < \theta$$

which is a contradiction. $\qquad\square$

LEMMA 1.6. *If a Banach space X is nonreflexive then there exists a decreasing sequence of closed and convex subsets $\{A_n\}$ in the unit ball $\overline{B}(0,1)$ such that*

$$0 < \inf\left\{\sup_{n \in \mathbb{N}} d(x, A_n) : x \in X\right\}.$$

Proof. If we suppose that

$$\inf\left\{\sup_{n\in N}\ d(x,A_n) : x\in X\right\}=0$$

for all decreasing sequences of closed and convex subsets $\{A_n\}$ in the unit ball $\overline{B}(0,1)$, then there exists $x_1 \in X$ such that $\sup_{n\in\mathbb{N}}\ d(x_1,C_n) < 1$, where $\{C_n\}$ is the sequence of the previous lemma. Let

$$A_n^1 = \{x\in C_n : \|x-x_1\| \le 1\}.$$

The sets A_n^1 are closed, convex and nonempty and the sequence $\{A_n^1\}$ is decreasing. So, there exists x_2 such that

$$\sup_{n\in\mathbb{N}}\ d(x_2,A_n^1) < \frac{1}{2}.$$

We define

$$A_n^2 = \{x\in A_n^1 : \|x-x_2\| \le \frac{1}{2}\}.$$

Inductively, suppose the sequence $\{A_n^{k-1}\}_{n\in\mathbb{N}}$ of closed, convex and nonempty sets has been constructed, then we choose x_k such that

$$\sup_{n\in\mathbb{N}}\ d(x_k,A_n^{k-1}) < \frac{1}{2^{k-1}}$$

and we define

$$A_n^k = \bar{B}\left(x_k,\frac{1}{2^{k-1}}\right)\bigcap A_n^{k-1}.$$

The sequence $\{A_n^k\}_{n\in\mathbb{N}}$ is decreasing and the sets are closed, convex and nonempty, with $A_n^{k+1} \subset A_n^k$ and $\text{diam}(A_n^k) \le 1/2^{k-2}$. If we consider the diagonal sequence $\{A_n^n\}$ we know that it is decreasing and the sets are closed, convex and nonempty with $\text{diam}(A_n^n) \to 0$. Thus by Cantor's theorem $\bigcap_{n=1}^\infty A_n^n \ne \emptyset$. But

$$\bigcap_{n=1}^\infty A_n^n \subset \bigcap_{n=1}^\infty C_n = \emptyset$$

which is a contradiction. $\qquad\square$

THEOREM 1.7. *If $\varepsilon_\chi(X) < 1$, then X is reflexive.*

Proof. If we suppose that X is nonreflexive then there exits a decreasing sequence A_n of closed and convex sets in the unit ball $\overline{B}(0,1)$ such that

$$0 < d = \inf\left\{\sup_{n \in \mathbb{N}} \, d(x, A_n) : x \in X\right\}.$$

Let $0 < \varepsilon < d$. There exists $z \in X$ such that

$$d - \varepsilon < d \leq \sup\{\, d(z, A_n) : n \in \mathbb{N}\} < d + \varepsilon.$$

As the sequence $\{\, d(z, A_n)\}$ is increasing then there exists n_0 such that if $n \geq n_0$ we have the following inequality

$$d - \varepsilon < \, d(z, A_n) < d + \varepsilon.$$

So for $n \geq n_0$ there exists x_n in A_n such that

$$d - \varepsilon < \, d(z, A_n) \leq \|z - x_n\| < d + \varepsilon.$$

Let $y_n = z - x_{n_0+n}$ and $A = \{y_n : n \in \mathbb{N}\}$. Then $A \subset B(0, d+\varepsilon)$ and for each $y \in X$ we have

$$\liminf_{n \to \infty} \|y_n - y\| = \liminf_{n \to \infty} \|z - x_n - y\| \geq \liminf_{n \to \infty} \, d(z - y, A_n) =$$
$$= \lim_{n \to \infty} \, d(z - y, A_n) = \sup\{\, d(z - y, A_n) : n \in \mathbb{N}\} \geq d.$$

So $\chi(A) \geq d$. Furthermore if $y \in \text{co}(\{y_n\})$ there exist $\lambda_1, \ldots, \lambda_n$ such that $\lambda_i \geq 0$, $\sum_{i=1}^{n} \lambda_i = 1$ and

$$\|y\| = \left\| z - \sum_{k=1}^{n} \lambda_k x_{n_0+k} \right\| \geq \, d(z, A_{n_0+1}) > d - \varepsilon.$$

Finally we consider $B = \dfrac{1}{d+\varepsilon} A$. Notice that $B \subset \overline{B}(0,1)$, $\chi(B) > d/d + 2\varepsilon$ and $d(0, \, \text{co}(B)) > (d - \varepsilon)/(d + \varepsilon)$. So

$$\Delta_\chi \left(\frac{d}{d + 2\varepsilon} \right) \leq 1 - \frac{d - \varepsilon}{d + \varepsilon}.$$

Since ε is an arbitrary positive number we have

$$\lim_{\varepsilon \to 0+} \Delta_\chi \left(\frac{d}{d + \varepsilon} \right) \leq 0.$$

Hence $\lim_{\varepsilon \to 1-} \Delta_\chi(\varepsilon) = 0$ and so $\varepsilon_\chi(X) \geq 1$ and the result follows. $\quad\square$

Remark 1.8. Bearing in mind the inequalities (2) in Remark 1.4 it is clear that X is reflexive if $\varepsilon_\phi(X) < 1$ for $\phi = \alpha$ or β. Furthermore Corollary 4.2.20 is a direct consequence of Theorem 1.7.

Now we are going to develop some alternative expressions for the moduli of noncompact convexity.

LEMMA 1.9. *Let X be a Banach space and $\{x_n\}$ a sequence in X weakly convergent to w. Let $A_n = \overline{co}(\{x_k\}_{k \geq n})$. Then*

$$\bigcap_{n=1}^{\infty} A_n = \{w\}.$$

Proof. We can suppose $w = 0$. Since $\{x_n\}_{n \geq k} \rightharpoonup 0$ then $0 \in A_n$ for all $n \in \mathbb{N}$ and so $0 \in \bigcap_{n=1}^{\infty} A_n$.

Let $a \in \bigcap_{n=1}^{\infty} A_n$, $a \neq 0$. There exists $f \in X^*$ such that $f(a) = \|a\|$ and $\|f\| = 1$. If $\varepsilon = \|a\|/3$, there exists n_0 such that $|f(x_n)| < \varepsilon$, $n \geq n_0$.

On the other hand, since $a \in A_{n_0}$, there exist $\lambda_1, \ldots, \lambda_k$ such that $\sum_{i=1}^{k} \lambda_i = 1$, $\lambda_i \geq 0$ and $\left\| a - \sum_{i=1}^{k} \lambda_i x_{n_0+i} \right\| < \varepsilon$. So

$$\|a\| = f(a) = f\left(a - \sum_{i=1}^{k} \lambda_i x_{n_0+i} \right) + f\left(\sum_{i=1}^{k} \lambda_i x_{n_0+i} \right)$$

$$\leq \varepsilon + \sum_{i=1}^{k} \lambda_i f(x_{n_0+i}) < 2\varepsilon < \|a\|$$

which is a contradiction. □

THEOREM 1.10. *Let X be a Banach space and P a property such that for every sequence in $\overline{B}(0,1)$ there exists a subsequence satisfying P. Then*

(a)

$$\Delta_{X,\beta}(\varepsilon) = \inf\{1 - d(\overline{co}(\{x_n\}), 0) : Sep(\{x_n\}) > \varepsilon, \{x_n\} \subset \overline{B}(0,1)$$
$$\text{and satisfying } P \}$$

where

$$Sep(\{x_n\}) = \sup\{\varepsilon : \|x_n - x_m\| \geq \varepsilon : n \neq m\}.$$

(b) *If, in addition, X is weakly compactly generated, then*

$$\Delta_{X,\chi}(\varepsilon) = \inf\{1 - d(\overline{co}(\{x_n\}), 0) : \chi(\{x_n\}) > \varepsilon, \{x_n\} \subset \overline{B}(0,1)$$
$$\text{and satisfying } P\}.$$

Proof. To prove (a) notice that $\beta(A) = \sup\{\varepsilon > 0 : \text{there exists a sequence } \{x_n\}$ in A such that $\text{Sep}(\{x_n\}) > \varepsilon\}$. It would be enough to prove $a = b$, where

$$a = \sup\left\{ \inf\{\|x\| : x \in \overline{co}(A)\} : \beta(A) > \varepsilon, A \subset \overline{B}(0,1)\right\},$$

$$b = \sup\left\{ \inf\{\|x\| : x \in \overline{co}(\{x_n\})\} : \text{Sep}(\{x_n\}) > \varepsilon, \{x_n\} \subset \overline{B}(0,1)\right.$$
$$\left. \text{and satisfying P}\right\}.$$

It is obvious that $a \geq b$.

On the other hand if $A \subset \overline{B}(0,1)$ with $\beta(A) > \varepsilon$, there exists a sequence $\{x_n\}$ such that $\text{Sep}(\{x_n\}) > \varepsilon, \{x_n\} \subset A$ and $\{x_n\}$ satisfies P. Thus

$$\inf\{\|x\| : x \in \overline{co}(\{x_n\})\} \geq \inf\{\|x\| : x \in \overline{co}(A)\}$$

which implies

$$\inf\{\|x\| : x \in co(A)\} \leq b.$$

As the subset A has been chosen arbitrarily, we conclude $a \leq b$.

For the proof of statement (b) we follow the same steps. In this case, to prove $a \geq b$, we use the fact that if the Banach space is weakly compactly generated, then from Theorems III.1.2 and III.2.7 we know that for every bounded subset A of X, there exists a subset B of A such that $\chi(B) = \chi(A)$ and B is χ-minimal. Taking a subset, if necessary, we can assume that B is a sequence and the proof runs as in the case (a). $\square$

THEOREM 1.11. *Let X be a reflexive Banach space and P a property such that for every sequence in $\overline{B}(0,1)$ there exists a subsequence satisfying P. Then*

$$\Delta_{X,\beta}(\varepsilon) = \inf\left\{1 - \|w\| : x_n \rightharpoonup w, \{x_n\} \subset \overline{B}(0,1), \{x_n\}\right.$$
$$\left. \text{satisfies P and } \text{Sep}(\{x_n\}) > \varepsilon\right\}$$

and

$$\Delta_{X,\chi}(\varepsilon) = \inf\left\{1 - \|w\| : x_n \rightharpoonup w, \{x_n\} \subset \overline{B}(0,1), \{x_n\}\right.$$
$$\left. \text{satisfies P and } \chi(\{x_n\}) > \varepsilon\right\}.$$

Proof. Both equalities have a similar proof. So we are only going to prove the first one. By the previous lemma

$$\Delta_{X,\beta}(\varepsilon) \geq \inf\left\{1 - \|w\| : x_n \rightharpoonup w, \{x_n\} \subset \overline{B}(0,1) \text{ and } \text{Sep}(\{x_n\}) > \varepsilon\right\}.$$

Conversely, let $x_n \rightharpoonup w$ with $\{x_n\} \subset \overline{B}(0,1)$ and $\text{Sep}(\{x_n\}) > \varepsilon$. As

$$1 - \Delta_{X,\beta}(\varepsilon) = \sup\left\{ d(0, \overline{co}(\{y_n\}) : \{y_n\} \subset \overline{B}(0,1) \ \text{Sep}(\{y_n\}) > \varepsilon\right\},$$

if we take $\{y_n\} = \{x_n\}_{n \geq k}$, we have

$$1 - \Delta_{X,\beta}(\varepsilon) \geq \mathrm{d}(0, \overline{\mathrm{co}}(\{x_n\}_{n \geq k})).$$

Let $\eta > 0$. For all k there exists

$$z_k \in \overline{\mathrm{co}}(\{x_n\}_{n \geq k}) = A_k$$

such that

$$\|z_k\| \leq 1 - \Delta_{X,\beta}(\varepsilon) + \eta.$$

Let z_{n_k} be a weakly convergent subsequence of $\{z_n\}$, and $B_n = \overline{\mathrm{co}}\{x_{n_k}\}_{k \geq n}$. Then $B_n \subset A_n$. So if $z_{n_k} \rightharpoonup z$ we obtain

$$z \in \bigcap_{n=1}^{\infty} B_n \subset \bigcap_{n=1}^{\infty} A_n = \{w\}.$$

Thus $z_n \rightharpoonup w$ and so

$$\|w\| \leq \liminf_{n \to \infty} \|z_n\| \leq 1 - \Delta_{X,\beta}(\varepsilon) + \eta.$$

Since the last inequality is true for every η, we obtain

$$\Delta_{X,\beta}(\varepsilon) \leq 1 - \|w\|$$

and again this inequality is true for every $x_n \rightharpoonup w$ satisfying $\{x_n\} \subset \overline{B}(0,1)$ and $\mathrm{Sep}(\{x_n\}) > \varepsilon$. So

$$\Delta_{X,\beta}(\varepsilon) \leq \inf\{1 - \|w\| : x_n \rightharpoonup w, (x_n) \subset \overline{B}(0,1) \ \ \mathrm{Sep}(\{x_n\}) > \varepsilon\}$$

and the proof is now complete. $\qquad\qquad\qquad\qquad\qquad\qquad\qquad\qquad\qquad\square$

Remark 1.12. The function

$$P_X(\varepsilon) = \inf\left\{1 - \|w\| : x_n \rightharpoonup w, \{x_n\} \subset \overline{B}(0,1) \text{ and } \ \mathrm{Sep}(\{x_n\}) > \varepsilon\right\}$$

has been considered by Partington [Pa] as a modulus for the UKK property. Theorem 1.10 shows that for reflexive spaces Partington's modulus is identical to the modulus of noncompact convexity associated to β.

Remark 1.13. By using the properties of the measures of noncompactness and Theorem 1.10, it is easy to deduce that a Banach space X is NUC if and only if $\varepsilon_\phi(X) = 0$, where ϕ is α, β or χ. From previous observations and Theorem 1.7 we may deduce that NUC implies reflexivity which, together with Theorem 1.11, directs us to the following conclusion.

THEOREM 1.14. *X is NUC if and only if X is UKK and reflexive.*

COROLLARY 1.15. *Every k-UC space is NUC.*

Proof. From Theorems IV.2.19 and IV.2.24 every k-UC space is UKK and reflexive. $\square$

We shall show in Example 2 an NUC space which is not k-UC for any $k \in \mathbb{N}$, but first we are going to evaluate the moduli of noncompact convexity of ℓ^p spaces.

THEOREM 1.16.

$$\Delta_{\ell^p,\beta}(\varepsilon) = 1 - \left(1 - \frac{\varepsilon^p}{2}\right)^{\frac{1}{p}}.$$

Proof. Let us see that $\Delta_{\ell^p,\beta}(\varepsilon) \leq 1 - (1 - \varepsilon^p/2)^{1/p}$. Indeed, if $\{e_n\}$ is the usual basis sequence in ℓ^p and $\eta > \varepsilon$ is arbitrary we consider the sequence $\{x_n\}$, where $x_n = (1 - \eta^p/2)^{1/p} e_1 + 2^{-1/p}\eta e_{n+1}$. Obviously $\|x_n\| = 1$, $\|x_n - x_m\| = \eta$ if $m \neq n$ and $\inf\{\|x\| : x \in \overline{co}(\{x_n\})\} = (1 - \eta^p/2)^{1/p}$. So

$$\Delta_{\ell^p,\beta}(\varepsilon) \leq 1 - \left(1 - \frac{\eta^p}{2}\right)^{\frac{1}{p}}$$

for all $\eta > \varepsilon$. Thus

$$\Delta_{\ell^p,\beta}(\varepsilon) \leq 1 - \left(1 - \frac{\varepsilon^p}{2}\right)^{\frac{1}{p}}.$$

Let us now study the converse inequality.

Since every ℓ^p, $1 < p < +\infty$, is reflexive we have

$$\Delta_{\ell^p,\beta}(\varepsilon) = 1 - \sup\{\inf \|x\| : x_n \rightharpoonup x, \text{ Sep}(\{x_n\}) \geq \varepsilon, \|x_n\| \leq 1\}.$$

Let $\{x_n\}$ be a sequence in ℓ^p such that $x_n \rightharpoonup x$, $\text{Sep}(\{x_n\}) > \varepsilon$, $\|x_n\| \leq 1$ and $\lim_{n\to\infty} \|x_n - x\| = l$.

Given $\eta > 0$, we can assume that $\|x_n - x\| \leq l + \eta$ for all $n \in \mathbb{N}$. A sufficiently large k can be chosen such that $\sum_{i>k} |x_1^i - x^i|^p < \eta^p$ and we denote $u = \sum_{i \leq k}(x_1^i - x^i)e_i$ and $v = \sum_{i>k}(x_1^i - x^i)e_i$. Then $x_1 - x = u + v$ and $\|v\| \leq \eta$. A sufficiently large n can be chosen such that $\sum_{i \leq k} |x_n^i - x^i|^p < \eta^p$ and we denote $u_n = \sum_{i \leq k}(x_n^i - x^i)e_i$ and $v_n = \sum_{i>k}(x_n^i - x^i)e_i$. Therefore $x_n - x = u_n + v_n$ and $\|u_n\| \leq \eta$.

The continuity of the function $t \longrightarrow t^p$ implies $\left||s+t|^p - |s|^p\right| < o(1)$ if $|t| < \eta$ where $o(1) \to 0$ as $\eta \to 0$. We now have

$$\begin{aligned}
\varepsilon^p &\leq \|x_1 - x_n\|^p = \|(x_1 - x) - (x_n - x)\|^p = \|u + v - u_n - v_n\|^p \\
&\leq \|u - v_n\|^p + o(1) = \|u\|^p + \|v_n\|^p + o(1) \\
&= \|x_1 - x - v\|^p + \|x_n - x - u_n\|^p + o(1) \\
&\leq \|x_1 - x\|^p + \|x_n - x\|^p + o(1) \leq 2l^p + o(1).
\end{aligned}$$

Since η is arbitrary we obtain $l \geq 2^{-1/p}\varepsilon$.

Again for any $\eta > 0$, a sufficiently large j can be chosen so that $\sum_{i>j} |x^i|^p < \eta^p$ and we denote $y = \sum_{i \leq j} x^i e_i$ and $z = \sum_{i>j} x^i e_i$. Therefore $x = y + z$ and $\|z\| \leq \eta$. For a large enough n we have $\sum_{i \leq j} |x^i - x_n^i|^p < \eta^p$. We denote $y_n = \sum_{i \leq j} (x_n^i - x^i) e_i$ and $z_n = \sum_{i>j} (x_n^i - x^i) e_i$. Therefore $x_n - x = y_n + z_n$ and $\|y_n\| < \eta$. We have

$$
\begin{aligned}
1 \geq \|x_n\|^p &= \|x_n - x + x\|^p \geq \|y_n + z_n + y + z\|^p \\
&= \|y_n + y\|^p + \|z_n + z\|^p \geq \|y\|^p + \|z_n\|^p - o(1) \\
&\geq \|x - z\|^p + \|x_n - x + y_n\|^p - o(1) \geq \|x\|^p + \|x_n - x\|^p - o(1).
\end{aligned}
$$

Thus $\|x\|^p \leq (1 - \dfrac{\varepsilon^p}{2}) + o(1)$, and letting $\eta \to 0$, we obtain

$$
\Delta_{\ell^p, \beta}(\varepsilon) \geq 1 - \left(1 - \frac{\varepsilon^p}{2}\right)^{\frac{1}{p}},
$$

that is, the required inequality. $\qquad\square$

Remark 1.17. The value of $\Delta_{\ell^p, \chi}(\varepsilon)$ is easily deduced by using the relationship between the measures of noncompactness β and χ in the spaces ℓ^p which will be given in Corollary X.4.7. This value is

$$
\Delta_{\ell^p, \chi}(\varepsilon) = 1 - (1 - (\varepsilon)^p)^{\frac{1}{p}}.
$$

Goebel and Sękowski [GS] computed the value of $\Delta_{\ell^p, \alpha}$ obtaining

$$
\Delta_{\ell^p, \alpha}(\varepsilon) = 1 - \left(1 - \left(\frac{\varepsilon}{2}\right)^p\right)^{\frac{1}{p}}
$$

which, for $p \geq 2$, coincides with the value of the Clarkson modulus of convexity (see Example IV.6).

Example 2: For $n \in \mathbb{N}$ let $\ell_n^1 = (\mathbb{R}^n, \|.\|_1)$ and $\ell_n^\infty = (\mathbb{R}^n, \|.\|_\infty)$. The following Banach spaces were defined by Day [D]:

$$
D_1 = \left\{ x = (x_i) : x_i \in \ell_n^1, \|x\|_{D_1} = \left(\sum_{i=1}^{\infty} (\|x_i\|_1)^2 \right)^{\frac{1}{2}} < \infty \right\}
$$

and

$$
D_\infty = \left\{ x = (x_i) : x_i \in \ell_n^\infty, \|x\|_{D_\infty} = \left(\sum_{i=1}^{\infty} (\|x_i\|_\infty)^2 \right)^{\frac{1}{2}} < \infty \right\}.
$$

If we take

$$e_1 = (1, 0, \ldots, 0), \ldots, e_k = (0, \ldots, 1) \in \ell_k^1$$

and

$$v_1 = (1, 0, \ldots, 0), \ldots, v_k = (1, \ldots, 1) \in \ell_k^\infty$$

it is easy to check that D_1 and D_∞ fail to be k-uniformly convex spaces, for any $k \in \mathbb{N}$.

We may compute the values of $\Delta_{D_1,\beta}(\varepsilon)$ and $\Delta_{D_\infty,\beta}(\varepsilon)$ in a similar way to the case of ℓ^2 obtaining

$$\Delta_{D_1,\beta}(\varepsilon) = \Delta_{D_\infty,\beta}(\varepsilon) = 1 - \left(1 - \frac{\varepsilon^2}{2}\right)^{\frac{1}{2}}.$$

So the spaces D_1 and D_∞ are NUC but fail to be k-UC for any k.

We now continue by studying some properties of the moduli of noncompactness. The continuity of $\Delta_{X,\chi}$ has been studied by Banaś [Ba1].

THEOREM 1.18. *The function $\Delta_{X,\chi}(\varepsilon)$ is continuous on the interval $[0, 1)$ for any Banach space X.*

Proof. Given $\varepsilon_1 \in [0, 1)$ choose $\varepsilon_2 \in [\varepsilon_1, 1)$.

From the definition of the modulus $\Delta_{X,\chi}(\varepsilon)$, given an arbitrarily small $\eta > 0$, we may choose a convex set X_1, contained in B_X such that, $\chi(X_1) > \varepsilon_1$ and

$$1 - d(0, X_1) \leq \Delta_{X,\chi}(\varepsilon_1) + \eta.$$

We consider the set $Y = kX_1$, where $k = (1 - \varepsilon_2)/(1 - \varepsilon_1)$. Notice that Y is convex, $\chi(Y) = k\chi(X_1)$, $d(0, Y) = k \, d(0, X_1)$ and for all $y \in Y$, $\|y\| \leq k$. So the set

$$X_2 = B(Y, 1 - k)$$

is a convex subset of the unit ball and

$$d(0, X_2) = k \, d(0, X_1) - 1 + k.$$

Moreover from Theorem II.2.10

$$\chi(X_2) = k\chi(X_1) + 1 - k > \varepsilon_2.$$

On the other hand

$$\begin{aligned}
1 - d(0, X_2) &= 1 - k \, d(0, X_1) + 1 - k = k(1 - d(0, X_1)) + 2(1 - k) \\
&\leq k(\Delta_{X,\chi}(\varepsilon_1) + \eta) + 2(1 - k).
\end{aligned}$$

Since η can be chosen arbitrarily small we have

$$\Delta_{X,\chi}(\varepsilon_2) \leq k\Delta_{X,\chi}(\varepsilon_1) + 2(1-k).$$

So

$$\Delta_{X,\chi}(\varepsilon_2) - \Delta_{X,\chi}(\varepsilon_1) \leq (1-k)(2 - \Delta_{X,\chi}(\varepsilon_1))$$
$$\leq 2(1-k) = 2\left(\frac{\varepsilon_2 - \varepsilon_1}{1 - \varepsilon_1}\right).$$

Thus the proof is complete. □

2. Nearly Uniformly Smooth Banach spaces

Dual properties of NUC have been defined and studied by different authors [Ba2], [Pr5]. They give different definitions and both call the property near uniform smoothness, NUS for short. Here we consider the definition of S. Prus and certain properties of NUS spaces which have been proved by the same author [Pr5].

DEFINITION 2.1. *A Banach space X is said to be nearly uniformly smooth (NUS) if for all $\varepsilon > 0$ there exists $\eta > 0$ such that for each t, $0 < t < \eta$, and for each basic sequence $\{x_n\}$ in $\overline{B}(0,1)$ there exists $k > 1$ such that*

$$\|x_1 + tx_k\| < 1 + \varepsilon t.$$

THEOREM 2.2. *Let X be a NUS Banach space. Then X is reflexive.*

Proof. Let X be NUS. Then there exists $\eta > 0$ such that for each positive $t < \min\{1, \eta\}$ and each basic sequence $\{x_n\}$ in $\overline{B}(0,1)$ there exists $k > 1$ such that

$$\|x_1 + tx_k\| < 1 + \frac{t}{2}.$$

Suppose X is not reflexive. By James's characterization of nonreflexivity (Lemma IV.2.19), if we take $\theta = 1 - \frac{t}{5} > 0$ there exists a norm one sequence $\{x_n\}$ in X and a sequence of linear functionals $\{f_n\}$ such that if $m > n$

$$\|x_n - x_m\| \geq |f_m(x_n - x_m)| = \theta = 1 - \frac{t}{5}$$

and

$$\left\|\frac{1}{2}(x_n + x_m)\right\| \geq f_n(\frac{1}{2}(x_n + x_m)) = \theta = 1 - \frac{t}{5}. \tag{3}$$

Since $\{x_n\}$ is not a Cauchy sequence, by taking a subsequence we may suppose that $\{x_n\}$ is a basic sequence (see [LT1, Remark in page 5]). However

$$\left\|\frac{1}{2}(x_1 + x_k)\right\| \leq \frac{1}{2}(\|x_1 + tx_k\| + \|(1-t)x_k\|) \leq 1 - \frac{t}{4}$$

which is a contradiction of (3). □

Now we formulate the main result of this section.

THEOREM 2.3. *Let X be a Banach space. Then*
  (a) *X is NUC if and only if X^* is NUS.*
  (b) *X is NUS if and only if X^* is NUC.*

Proof. The statement (b) of the theorem is a direct consequence of statement (a) and Theorem 2.2.

Let us assume that X is NUC. Then by Theorem 1.14, X is UKK and reflexive. Therefore for a given ε, $0 < \varepsilon < 2$, there exists $\delta > 0$ such that if $\{x_n\}$ is a sequence in the unit ball $\overline{B}(0,1)$ with $\mathrm{Sep}(\{x_n\}) > \varepsilon/4$ and weakly convergent to x, then $\|x\| < 1 - \delta$.

We set $\eta = \delta/(2 - \varepsilon) > 0$ and let t be a fixed positive number less than η. If $\{x_n^*\}$ is a basic sequence in the unit ball of X^*, then by the reflexivity of X there exists a sequence $\{x_n\}$ in the unit ball of X so that

$$\|x_1^* + tx_n^*\| = (x_1^* + tx_n^*)(x_n).$$

Moreover, by taking a subsequence, we can assume that $\{x_n\}$ converges weakly to some $x \in X$.

From the reflexivity of X^* it follows that $\{x_n^*\}$ is shrinking (see [B, Theorem 5, page 86]) and so converges weakly to zero. Hence given

$$\gamma = \min\left\{\delta - t(2 - \varepsilon), \frac{1}{2}t\varepsilon\right\} > 0$$

there exists k_0 such that $x_k^*(x) < \gamma/2t$ and $x_1^*(x_k - x) < \gamma/2$ for all $k > k_0$. Therefore we have

$$\|x_1^* + tx_k^*\| = (x_1^* + tx_k^*)(x_k) \le x_1^*(x) + tx_k^*(x_k - x) + \gamma \tag{4}$$
$$\le \|x\| + t\|x_k - x\| + \gamma.$$

Let us consider two cases:

(i) $\|x_k - x\| < \varepsilon/2$ for certain $k > k_0$. In this case the result comes directly from (4) since

$$\|x_1^* + tx_k^*\| \le \|x\| + t\|x_k - x\| + \gamma$$
$$\le 1 + t\|x_k - x\| + \gamma \le 1 + \varepsilon t.$$

(ii) $\|x_k - x\| \ge \varepsilon/2$ for all $k > k_0$. Since $\{x_k - x\}$ converges weakly to zero and $\liminf_{n \to \infty} \|x_k - x\| > 0$, we can choose an increasing sequence of integers $\{n_k\}$, so that $\{x_{n_k} - x\}$ is a basic sequence with the basic constant arbitrarily close to 1, in particular less than 2 [LT1, Remark and proof of Theorem 1.a.5, page 4]. Thus

$$\frac{\varepsilon}{2} \le \|x_{n_1} - x\| \le 2\|(x_{n_k} - x) - (x_{n_1} - x)\| = 2\|x_{n_k} - x_{n_1}\|.$$

Therefore by our assumption $\|x\| \leq 1 - \delta$. Using (4) again we obtain

$$\|x_1^* + tx_k^*\| \leq \|x\| + t\|x_k - x\| + \gamma$$
$$\leq 1 - \delta + t\|x_k - x\| + \gamma \leq 1 - \delta + 2t + \gamma \leq 1 + \varepsilon t$$

for certain $k > k_0$. Now we may deduce that X^* is NUS.

Conversely we now suppose that X^* is NUS. Then by Theorem 2.2, X^* is reflexive and it suffices to prove that X is UKK. From our assumption it follows that for a given $\varepsilon > 0$ there exists $t > 0$ such that if $\{x_n^*\}$ is a basic sequence in the unit ball of X^*, then there exits k such that

$$\|x_1^* + tx_k^*\| \leq 1 + \frac{\varepsilon}{32}t. \tag{5}$$

Let $\{x_n\}$ be a sequence in the unit ball of X such that $\|x_n - x_m\| > \varepsilon$ for $n \neq m$ and $\{x_n\}$ is weakly convergent to x. We may suppose $x \neq 0$ and that $\|x_n - x\| > \varepsilon/2$ for all n. Then there exists a basic sequence $\{y_n\}$ with the basic constant $K \leq (1 + t\varepsilon/16)(1 + t\varepsilon/32)^{-1}$ such that $y_1 = x$ and $\{y_n\}_{n>1}$ is a subsequence of $\{x_n - x\}$.

The linear functionals z_n^* defined by $z_n^*\left(\sum_{i=1}^{\infty} a_i y_i / \|y_i\|\right) = a_n$ are defined on the space generated by $\{y_m/\|y_m\|\}$ and, with the help of the Hahn Banach theorem, we obtain a sequence $\{z_n^*\}$ of elements of X^* such that $z_n^* (y_m/\|y_m\|) = \delta_n^m$, $\|z_1^*\| \leq K$ and $\|z_n^*\| \leq 2K$ for all $n \geq 2$. Again by taking a subsequence we may assume that $\{z_n^*\}$ is weakly convergent to z^* where $z^*(y_n) = 0$. Therefore

$$\|z_n^* - z^*\| \geq \frac{1}{2} z_n^*(y_n) \geq \frac{\varepsilon}{4}$$

for all $n > 1$ and there is a basic sequence $\{x_n^*\}$ such that $x_1^* = z_1^*$ and $x_k^* = z_{n_k}^* - z^*$ for certain n_k if $k > 1$. Therefore we obtain

$$\|x\| = x_1^*(y_1) = x_1^*(y_1 + y_{n_k}) + \frac{t}{4} x_k^*(y_1 + y_{n_k}) - \frac{t}{4} x_k^*(y_1 + y_{n_k}) =$$
$$= (x_1^* + \frac{t}{4} x_k^*)(y_1 + y_{n_k}) - \frac{t}{4} z_{n_k}^*(y_{n_k}) \leq \left\|x_1^* + \frac{t}{4} x_k^*\right\| - \frac{\varepsilon}{8} t$$

for all $k > 1$.

Since $\|x_1^*\| \leq K$, $\frac{1}{4}\|x_k^*\| \leq K$ by (5) we have

$$\left\|x_1^* + \frac{t}{4} x_k^*\right\| \leq K\left(1 + \frac{t\varepsilon}{32}\right) \leq 1 + \frac{t\varepsilon}{16}$$

for certain k.

So we may conclude that there exists $\delta = (\varepsilon/16)t > 0$ which only depends on ε such that $\|x\| \leq 1 - \delta$. Hence the space X is UKK. $\qquad \square$

As a consequence of the Theorems 2.3 and IV.3.3 we have the following result:

COROLLARY 2.4. *If X is k-US for some k, then X is NUS.*

A modulus of near uniform smoothness has been defined in [Do7].

Definition 2.5. *Let X be a Banach space. We define the modulus of near uniform smoothness of X as the function*

$$\Gamma_X(t) = \sup\left\{\inf\left\{\frac{\|x_1 + tx_n\| + \|x_1 - tx_n\|}{2} - 1 : n > 1\right\} : \{x_n\}\right.$$

$$\left. \text{basic sequence in } \overline{B}(0,1)\right\}.$$

It is obvious from the definition that $\Gamma_X(t) \le \rho_X(t)$ for every $t \ge 0$. Thus if X is US we obtain $\lim_{t\to 0^+} \Gamma_X(t)/t = 0$. We shall give an equivalent definition for $\Gamma_X(t)$ when X is a reflexive space.

THEOREM 2.6. *Let X be a reflexive Banach space. Then*

$$\Gamma_X(t) = \sup\left\{\inf\left\{\frac{\|x_1 + tx_n\| + \|x_1 - tx_n\|}{2} - 1 : n > 1\right\} : \{x_n\}\right.$$

$$\left. \text{weakly null in } \overline{B}(0,1)\right\}.$$

Proof. Let

$$\hat{\Gamma}_X(t) = \sup\left\{\inf\left\{\frac{\|x_1 + tx_n\| + \|x_1 - tx_n\|}{2} - 1 : n > 1\right\} : \{x_n\}\right.$$

$$\left. \text{weakly null in } \overline{B}(0,1)\right\}.$$

Since X is reflexive, every basic sequence is weakly null. Thus $\hat{\Gamma}_X(t) \ge \Gamma_X(t)$. On the other hand, let $\{x_n\}$ be a weakly null sequence in $\overline{B}(0,1)$. If $\liminf_{n\to\infty} \|x_n\| > 0$, there exists a basic subsequence $\{y_n\}$ of $\{x_n\}$ such that $y_1 = x_1$. Thus

$$\Gamma_X(t) \ge \inf\left\{\frac{\|y_1 + ty_n\| + \|y_1 - ty_n\|}{2} - 1 : n > 1\right\}$$

$$\ge \inf\left\{\frac{\|x_1 + tx_n\| + \|x_1 - tx_n\|}{2} - 1 : n > 1\right\}. \tag{6}$$

If $\liminf_{n\to\infty} \|x_n\| = 0$, there exists a subsequence $\{y_n\}$ of $\{x_n\}$ such that $y_1 = x_1$ and $\lim_{n\to\infty} y_n = 0$. Let η be an arbitrary number bigger than $\Gamma_X(t)$. There exists an integer n_0 such that $\|y_n\| < \eta/t$ if $n \ge n_0$. Then

$$\eta \ge \frac{\|y_1 + ty_n\| + \|y_1 - ty_n\|}{2} - 1 \ge \inf\left\{\frac{\|y_1 + ty_n\| + \|y_1 - ty_n\|}{2} - 1 : n > 1\right\}$$

$$\ge \inf\left\{\frac{\|x_1 + tx_n\| + \|x_1 - tx_n\|}{2} - 1 : n > 1\right\}. \tag{7}$$

From (6) and (7) we obtain $\eta \ge \hat{\Gamma}_X(t)$. Thus $\hat{\Gamma}_X(t) \le \Gamma_X(t)$. $\qquad\square$

THEOREM 2.7. *Let X be a Banach space. Then X is NUS if and only if X is reflexive and*

$$\lim_{t \to 0^+} \frac{\Gamma_X(t)}{t} = 0.$$

Proof. If $\lim_{t \to 0^+} \Gamma_X(t)/t = 0$, for every $\varepsilon > 0$ there exists $\eta > 0$ such that $\Gamma_X(t) \le t\varepsilon$ for $t \in [0, \eta]$. Let $\{x_n\}$ be a basic sequence in $\overline{B}(0, 1)$. Since X is reflexive, $\{x_n\}$ is weakly null and we can assume that $\{x_n\}$ is not norm convergent, otherwise the proof is direct. Using a similar argument to that in the proof of Theorem 2.3 we can construct a subsequence $\{x_{n_k}\}$ such that $x_{n_1} = x_1$ and $\{x_{n_k}\}$ is a basic sequence with constant $c > 1$, where $1 + c < (1 + 3t\varepsilon)/(1 + 2t\varepsilon)$. So

$$\|x_1 + tx_{n_k}\| \le \frac{1}{2}(\|x_1\| + \|x_1 + 2tx_{n_k}\|)$$

$$\le \frac{1}{2}((1 + c)\|x_1 - 2tx_{n_k}\| + \|x_1 + 2tx_{n_k}\|).$$

Then for some k we have

$$\|x_1 + tx_{n_k}\| \le (1 + c)(1 + 2t\varepsilon) < 1 + 3t\varepsilon.$$

Conversely, if X is NUS then X is reflexive. Let $\{x_n\}$ be a weakly null sequence. Then for every $\varepsilon > 0$ there exists $\eta > 0$ such that $\|x_1 + tz_n\| \le 1 + \varepsilon t$ for every $n > 1$, where $\{z_n\}$ is a subsequence of $\{x_n\}$ with $z_1 = x_1$. Since the sequence $\{x_1, -z_2, -z_3, \dots\}$ is also weakly null we have $\|x_1 - tz_n\| \le 1 + \varepsilon t$ for some $n > 1$ and every $t \in [0, \eta]$. Thus

$$\frac{1}{2}(\|x_1 + tz_n\| + \|x_1 - tz_n\|) - 1 \le \varepsilon t$$

if $t \in [0, \eta]$. Hence $\lim_{t \to 0^+} \Gamma_X(t)/t = 0$. $\qquad\square$

Example 3: We know that

$$\Gamma_{\ell^2} \le \rho_{\ell^2}(t) = \sqrt{1 + t^2} - 1.$$

On the other hand, if $\{e_n\}$ is the usual basis sequence in ℓ^2, since

$$\frac{\|e_1 + te_k\| + \|e_1 - te_k\|}{2} - 1 = \sqrt{1 + t^2} - 1,$$

we have

$$\Gamma_{\ell^2}(t) \ge \sqrt{1 + t^2} - 1.$$

Thus

$$\Gamma_{\ell^2}(t) = \sqrt{1 + t^2} - 1.$$

3. Uniform Opial condition

DEFINITION 3.1. *A Banach space X has the Opial property if for every weakly null sequence $\{x_n\}$ and every $x \neq 0$ in X*

$$\liminf_{n \to \infty} \|x_n\| < \liminf_{n \to \infty} \|x + x_n\|.$$

In Lemma III.2.3 we have proved that Hilbert spaces have this property. In this section we will prove that ℓ^p, $1 < p < \infty$ also have the property. The following example, due to Opial [O1], shows that $L^p[0, 2\pi]$ has not the Opial property for $p \neq 2$.

Example 4: Let ϕ be a periodic real valued function of period 2π such that

$$\phi(t) = \begin{cases} 1 & \text{if } 0 \leq t \leq \frac{4}{3}\pi \\ -2 & \text{if } \frac{4}{3}\pi < t < 2\pi. \end{cases}$$

Consider the sequence $\{\varphi_n\}$, where $\varphi_n(t) = \phi(nt)$, in $L^p[0, 2\pi]$. First we are going to show that $\varphi_n \rightharpoonup 0$. Indeed, the step functions are dense in $L^q[0, 2\pi]$ and if ψ is a step function we have

$$\lim_{n \to \infty} \int_0^{2\pi} \phi(nt)\psi(t)dt = 0.$$

Now we are going to show that there exists a function $\psi \neq 0$ in $L^p[0, 2\pi], p \neq 2$, such that

$$\liminf_{n \to \infty} \|\varphi_n\| < \liminf_{n \to \infty} \|\psi + \varphi_n\|.$$

For any constant c, if we consider the constant function $\psi(t) \equiv c$, and we define

$$\Phi_p(c) = \lim_{n \to \infty} \|\varphi_n - c\|_p^p = \int_0^{2\pi} |\phi(t) - c|^p dt,$$

we have

$$\Phi_p'(0) = -p \int_0^{2\pi} |\phi(t)|^{p-1} \operatorname{sgn}(\phi(t))dt.$$

By the definition of ϕ, $\Phi_p'(0) \neq 0$ whenever $p \neq 2$. This implies that $\Phi_p(0)$ is not an extreme value of the function $\Phi_p(c)$, except for the case $p = 2$.

In 1992, Prus [Pr2] introduced the notion of the uniform Opial condition.

DEFINITION 3.2. *A Banach space X is said to satisfy the uniform Opial condition if for every $c > 0$, there exists an $r = r(c) > 0$ such that*

$$1 + r \leq \liminf_{n \to \infty} \|x + x_n\|$$

for all $x \in X$ with $\|x\| \geq c$ and sequences $\{x_n\}$ in X such that $\{x_n\}$ weakly converges to 0 and $\liminf_{n \to \infty} \|x_n\| \geq 1$.

The following modulus associated to the uniform Opial condition has been defined in [LTX].

DEFINITION 3.3. *Let X be a Banach space. The modulus of Opial associated to the space X, denoted by $r_X(c)$, is defined for $c \geq 0$ to be*

$$r_X(c) = \inf\{\liminf_{n \to \infty} \|x + x_n\| - 1\}, \quad c \geq 0,$$

where the infimum is taken over all $x \in X$ with $\|x\| \geq c$ and all weakly null sequences $\{x_n\}$ in X with $\liminf_{n \to \infty} \|x_n\| \geq 1$.

Remark 3.4. It is easily seen that the uniform Opial condition implies Opial's condition and that X satisfies the uniform Opial condition if and only if $r_X(c) > 0$ for all $c > 0$.

THEOREM 3.5. *The modulus of Opial associated to a Banach space satisfies the following properties:*
  (a) *r_X is non-decreasing.*
  (b) *$r_X(c) \leq c$ for all $c > 0$.*
  (c) *$r_X(c_2) - r_X(c_1) \leq c_2(c_2 - c_1)/c_1$ for all $c_2 \geq c_1 > 0$.*
  (d) *If $r_X(0) < 0$, then r_X is constant in $[0, -r_X(0)]$.*
  (e) *r_X is continuous in $[0, +\infty)$.*

Proof.
(a) It is obvious from the definition.

(b) $\qquad r_X(c) \leq \inf\{c + \liminf_{n \to \infty} \|x_n\| - 1 : \liminf_{n \to \infty} \|x_n\| \geq 1, \{x_n\} \rightharpoonup 0\} = c.$

(c) Let $c_2 > c_1 > 0$ and $\varepsilon > 0$ be arbitrary. By definition of Opial's modulus there exists $\{x_n\} \rightharpoonup 0$, $\liminf_{n \to \infty} \|x_n\| \geq 1$, $\|x\| \geq c_1$ such that

$$\liminf_{n \to \infty} \|x_n + x\| \leq 1 + r_X(c_1) + \varepsilon.$$

Let us consider two cases:

(i) If $\|x\| \geq c_2$ we would have

$$r_X(c_2) \leq r_X(c_1) + \varepsilon \leq r_X(c_1) + \varepsilon + \frac{c_2}{c_1}(c_2 - c_1)$$

and the proof would be concluded.

(ii) If $\|x\| \leq c_2$ we let $y = \frac{c_2}{c_1}x$. Since $\|y\| \geq c_2$ we have

$$\liminf_{n\to\infty} \|x_n + y\| \leq \liminf_{n\to\infty} \|x_n + x\| + \|x\|(\frac{c_2}{c_1} - 1)$$

$$\leq 1 + r_X(c_1) + \varepsilon + \frac{c_2}{c_1}(c_2 - c_1).$$

So

$$r_X(c_2) \leq r_X(c_1) + \varepsilon + \frac{c_2}{c_1}(c_2 - c_1)$$

and the proof is concluded.

(d) Let $a > 1$ and $0 < \varepsilon < -r_X(0)/a$. There exists $\{x_n\} \rightharpoonup 0$, $\liminf_{n\to\infty} \|x_n\| \geq 1$, and $x \in X$ such that

$$\liminf_{n\to\infty} \|x_n + x\| \leq 1 + r_X(0) + \varepsilon. \tag{8}$$

So

$$1 - \|x\| \leq \liminf_{n\to\infty} \|x_n\| - \|x\| \leq \liminf_{n\to\infty} \|x + x_n\| \leq 1 + r_X(0) + \varepsilon.$$

Thus

$$\|x\| > -r_X(0)\left(1 - \frac{1}{a}\right)$$

and then from (8)

$$r_X\left(-r_X(0)\left(1 - \frac{1}{a}\right)\right) \leq r_X(0) + \varepsilon.$$

Since ε is arbitrary

$$r_X\left(-r_X(0)\left(1 - \frac{1}{a}\right)\right) \leq r_X(0).$$

Notice that from statement (c) we may deduce that $r_X(c)$ is continuous in $(0, +\infty)$. Since a is arbitrary and $-r_X(0)(1 - \frac{1}{a}) > 0$ we have

$$r_X(-r_X(0)) \leq r_X(0)$$

and so $r_X(c)$ is constant in $[0, -r_X(0)]$.

(e) It is easily deduced from (b), (c) and (d). $\qquad\square$

Given a continuous strictly increasing function $\varphi : \mathbb{R}^+ \to \mathbb{R}^+$ such that $\varphi(0) = 0$ and $\lim_{r\to\infty} \varphi(r) = \infty$, we associate a (possibly multivalued) *generalized duality map* $J_\varphi : X \to \mathcal{P}(X^*)$, defined as

$$J_\varphi(x) = \{x^* \in X^* : x^*(x) = \|x\|\varphi(\|x\|) \text{ and } \|x^*\| = \varphi(\|x\|)\}$$

for every $x \in X$.

DEFINITION 3.6. *A space X is said to have a weakly continuous generalized duality map if there exists a continuous strictly increasing function $\varphi : \mathbb{R}^+ \to \mathbb{R}^+$ such that $\varphi(0) = 0$ and $\lim_{r\to\infty} \varphi(r) = \infty$ and the generalized duality map J_φ is single-valued and (sequentially) continuous from X with the weak topology to X^* with the weak* topology.*

Every ℓ^p-space ($1 < p < \infty$) has a weakly continuous generalized duality map for $\varphi(t) = t^{p-1}$ (see, for instance, [Br4, pag 112]).

Gossez and Lami Dozo [GL] proved that a space with a weakly continuous duality map satisfies Opial's condition. Lin, Tan and Xu [LTX] proved that a space X with a weakly continuous duality map must satisfy the uniform Opial condition. To prove this result we need a lemma due to Gossez and Lami Dozo [GL].

LEMMA 3.7. *If J_φ is single valued, then*

$$\Phi(\|x + y\|) = \Phi(\|x\|) + \int_0^1 J_\varphi(x + ty)(y)dt,$$

where

$$\Phi(t) = \int_0^t \varphi(s)ds.$$

We have not included the proof of this lemma because it uses some technical results which are outside our interest.

THEOREM 3.8. *Suppose X is a Banach space with a weakly continuous duality map. Then*

$$r_X(c) = \Phi^{-1}(\Phi(1) + \Phi(c)) - 1, \quad c \geq 0.$$

Proof. Suppose that $\{x_n\}$ is a sequence in X converging weakly to x. Then, considering the above lemma and that J_φ is weakly continuous, we have

$$\liminf_{n\to\infty} \Phi(\|x_n + y\|) = \liminf_{n\to\infty} \Phi(\|x_n\|) + \int_0^1 J_\varphi(x + ty)(y)dt.$$

In particular, if $\{x_n\}$ is weakly convergent to 0 and $\liminf_{n\to\infty} \|x_n\| \geq 1$, then

$$\liminf_{n\to\infty} \Phi(\|x_n + y\|) \geq \Phi(1) + \int_0^1 J_\varphi(ty)(y)dt$$

$$= \Phi(1) + \|y\| \int_0^1 \varphi(\|ty\|)dt$$

$$= \Phi(1) + \Phi(\|y\|).$$

Thus for all $y \in X$ we obtain

$$\liminf_{n\to\infty} \|x_n + y\| \geq \Phi^{-1}(\Phi(1) + \Phi(\|y\|))$$

which by definition of Opial's modulus implies

$$r_X(c) \geq \Phi^{-1}(\Phi(1) + \Phi(c)) - 1, \quad c \geq 0.$$

In order to obtain the other inequality let us notice that if $\{x_n\}$ is a sequence of norm one vectors in X converging weakly to 0, and y a point of X with norm c, then

$$r_X(c) \leq \liminf \|x_n + y\| - 1 = \Phi^{-1}(\Phi(1) + \Phi(c)) - 1, \quad c \geq 0.$$

$\square$

Remark 3.9. Theorem 3.8 lets us compute Opial's modulus of ℓ^p, obtaining

$$r_{\ell^p}(c) = (1 + c^p)^{\frac{1}{p}} - 1.$$

So ℓ^p-spaces, $(1 < p < \infty)$, satisfy the uniform Opial condition.

COROLLARY 3.10. *If a Banach space has a weakly continuous duality map then it satisfies the uniform Opial condition.*

A more precise connection between uniform opial property and duality mapping has been studied in [DaS].

Now we present an example of a Banach space that satisfies the uniform Opial condition but fails to have a weakly continuous duality map.

Example 5: We are going to show that if $1 < p < \infty$, then the space $\ell^{p,1}$ satisfies the uniform Opial condition, with Opial's modulus

$$r_{\ell^{p,1}} = (1 + c^p)^{\frac{1}{p}} - 1, \quad c \geq 0.$$

Indeed, let $f(t_1, t_2, t_3, t_4)$ be the function $(t_1^p + t_3^p)^{\frac{1}{p}} + (t_2^p + t_4^p)^{\frac{1}{p}}$. Then it is easily seen that for any $c > 0$, the infimum of f over the domain

$$D_c = \{(t_1, t_2, t_3, t_4) : 0 \leq t_1, t_2, t_3, t_4, \quad t_1 + t_2 \geq 1 \quad \text{and} \quad t_3 + t_4 \geq c\}$$

is achieved at the point $(\frac{1}{2}, \frac{1}{2}, \frac{c}{2}, \frac{c}{2})$ with value $(1 + c^p)^{\frac{1}{p}}$. Now suppose $\{x_n\}$ is a sequence in $\ell^{p,1}$ such that $\{x_n\}$ weakly converges to 0 and $\liminf_{n \to \infty} \|x_n\|_{p,1} \geq 1$, and x is an element of X with at least norm c. We choose a subsequence $\{x_{n_k}\}$ of $\{x_n\}$ such that

$$\lim_{k \to \infty} \|x_{n_k} + x\|_{p,1} = \liminf_{n \to \infty} \|x_n + x\|_{p,1}$$

and where $\lim_{k \to \infty} \|x_{n_k}\|_{p,1}$, $B = \lim_{k \to \infty} \|x_{n_k}^+\|_p$ and $C = \lim_k \|x_{n_k}^-\|_p$ exist. Since

$$\lim_{k \to \infty} \|x_{n_k}\|_{p,1} = \lim_{k \to \infty} (\|x_{n_k}^+\|_p + \|x_{n_k}^-\|) \geq \liminf_{n \to \infty} \|x_n\|_{p,1} \geq 1,$$

we have $B + C \geq 1$. Since $\{x_{n_k}\}$ is weakly null, by using standard arguments we can find sequences $\{y_k\}$, $\{z_k\}$, $\{u_k\}$ and $\{v_k\}$ such that $x = y_k + z_k$, $x_{n_k} = u_k + v_k$ where $\lim_{k\to\infty} z_k = \lim_{k\to\infty} u_k = 0$, $\operatorname{supp}(y_k) \cap \operatorname{supp}(v_k) = \emptyset$. Therefore

$$
\begin{aligned}
\liminf_{n\to\infty} \|x_n + x\|_{p,1} &= \lim_{k\to\infty} \|x_{n_k} + x\|_{p,1} \\
&= \lim_{k\to\infty} \|v_k + y_k\|_{p,1} = \lim_{k\to\infty} (\|v_k^+ + y_k^+\|_p + \|v_k^- + y_k^-\|_p) \\
&= (B^p + \|x^+\|_p^p)^{\frac{1}{p}} + (C^p + \|x^-\|_p^p)^{\frac{1}{p}} \\
&= f(B, C, \|x^+\|_p, \|x^-\|_p) \\
&\geq \inf_{D_c} f(t_1, t_2, t_3, t_4) \\
&= (1 + c^p)^{\frac{1}{p}}.
\end{aligned}
$$

It follows that $r_{\ell^{p,1}}(c) \geq (1 + c^p)^{\frac{1}{p}} - 1$. By considering the sequence $\{e_n\}$, we conclude that $r_{\ell^{p,1}}(c) = (1 + c^p)^{\frac{1}{p}} - 1$.

Considering the elements $x = e_1 - e_2$ and $y = e_1 - e_3$, we see that the dual space $\ell^{p,\infty}$ of $\ell^{p,1}$ is not strictly convex and hence $\ell^{p,1}$ is not smooth and so any generalized duality mapping is multivalued. Therefore, $\ell^{p,1}$ demonstrates a class of Banach spaces which satisfies the uniform Opial condition but fails to have a weakly continuous duality map.

We finish this chapter with a result due to Prus [Pr2] which establishes a relationship between the uniform Opial condition and the modulus $\Delta_{X,\chi}(\varepsilon)$.

THEOREM 3.11. *Let X be a Banach space. Then $\Delta_{X,\chi}(1^-) = 1$ if and only if X is reflexive and has the uniform Opial condition.*

Proof. Assume $\Delta_{X,\chi}(1^-) = 1$. Then X is reflexive. We will now prove that X has the uniform Opial condition.

Let $\eta > 0$ be arbitrary. There exists a weakly null sequence $\{x_n\}$ with $\liminf_{n\to\infty} \|x_n\| \geq 1$ and a vector x, $\|x\| \geq c$ such that

$$
\liminf_{n\to\infty} \|x + x_n\| < 1 + r_X(c) + \eta.
$$

Furthermore the definition of $r_X(0)$ implies $\liminf_{n\to\infty} \|y + x_n\| \geq 1 + r_X(0)$ for every $y \in X$. Thus

$$
\chi(\{x + x_n\}) = \chi(\{x_n\}) \geq 1 + r_X(0).
$$

Let

$$
y_n = \frac{x_n + x}{1 + r_X(c) + \frac{\eta}{2}}.
$$

Therefore

$$
\chi(\{y_n\}) \geq \frac{1 + r_X(0)}{1 + r_X(c) + \eta}
$$

and $y_n \in \overline{B}(0,1)$ for a large enough n. Since

$$y_n \to \frac{x}{1 + r_X(c) + \frac{\eta}{2}}$$

we obtain

$$\frac{c}{1 + r_X(c) + \frac{\eta}{2}} \leq \frac{\|x\|}{1 + r_X(c) + \frac{\eta}{2}} \leq 1 - \Delta_{X,x}\left(\frac{1 + r_X(0)}{1 + r_X(c) + \eta}\right).$$

Since $\eta > 0$ is arbitrary we conclude

$$\frac{c}{1 + r_X(c)} \leq 1 - \Delta_{X,x}\left(\left(\frac{1 + r_X(0)}{1 + r_X(c)}\right)^-\right)$$

and so X has the uniform Opial condition, because if for some c_0, $r_X(c_0) = r_X(0)$ then $c_0 = 0$. Since r_X is constant on $[0, -r_X(0)]$ we see that $r_X(0) = 0$ and hence $r_X(c) > 0$ for every $c > 0$.

Conversely, we assume that X is reflexive and has the uniform Opial property. We choose an arbitrary positive number $\varepsilon < 1$. For any $\eta > 0$ we may find a sequence in $\overline{B}(0,1)$ such that $\chi(\{x_n\}) > \varepsilon$, $x_n \to w$ and $\|w\| \geq 1 - \eta - \Delta_{X,x}(\varepsilon)$. Let $y_n = x_n - w$. We may assume that $\{x_n\}$ is χ-minimal because X is reflexive (see Theorem III.2.7). Then $\liminf_{n\to\infty} \|x_n - x\| > \varepsilon$ for every $x \in X$. Hence

$$\frac{1}{\varepsilon} \geq \frac{\|x_n\|}{\varepsilon} = \frac{\|y_n + w\|}{\varepsilon}$$

$$\geq 1 + r_X\left(\frac{\|w\|}{\varepsilon}\right) \geq 1 + r_X\left(\frac{1 - \eta - \Delta_{X,x}(\varepsilon)}{\varepsilon}\right).$$

If $1 - \eta - \Delta_{X,x}(1^-)$ were a positive number, taking limits as $\varepsilon \to 1^-$ and using the continuity of r_X we would obtain $r_X(1 - \eta - \Delta_{X,x}(1^-)) \leq 0$ which would be a contradiction because X has the uniform Opial condition. Thus $1 - \eta - \Delta_{X,x}(1^-) \leq 0$ and since η is arbitrary we conclude that $\Delta_{X,x}(1^-) = 1$. $\square$

Chapter VI

Fixed Points for Nonexpansive Mappings and Normal Structure

The most known and important metric fixed point theorem is the Banach fixed point theorem, also called the *contractive mapping principle*, which assures that every contraction from a complete metric space into itself has a unique fixed point. We recall that a mapping T from a metric space (X, d) into itself is said to be a *contraction* if there exists $k \in [0, 1)$ such that $d(Tx, Ty) \leq kd(x, y)$ for every $x, y \in X$. This theorem appeared in explicit form in Banach's Thesis in 1922 [Bn] where it was used to establish the existence of a solution for an integral equation. The simplicity of its proof and the possibility of attaining the fixed point by using successive approximations have made this theorem a very useful tool in Analysis and in Applied Mathematics.

> **Stefan Banach** (1892–1945) was born in Cracow (Poland) on 30 March. The birth certificate states that Banach's mother was Katarzyna Banach and that his father was Stefan Greczek, a civil servant. They were not married. Banach never knew his mother, who gave him up after his baptism. On many occasions Banach tried to learn something about her from his father, but the mystery was never solved since his father refused to reveal her identity or divulge any information whatsoever on the subject. Though Banach's father established his own legitimate family, he never forgot his son. Not only did he often provide some financial help, but he maintained close contact with his educators.
>
> At first he studied mathematics on his own. Afterwards he entered Lwów Technical University. His studies at the Technical University were interrupted by the outbreak of the World War I. However, he did not cease to be interested in mathematics, though he followed no formal course of study at that time.
>
> In 1920, Professor Lomnicki engaged Banach as his assistant at the Lwów Technical University although Banach had not yet finished his studies. That was the beginning of Banach's meteoric scientific career. In the same year, Banach submitted his doctor's dissertation to Jan

Kazimierz University in Lwów. It was published in the third volume of "Fundamenta Mathematica" under the title of "Sur les opérations dans les ensembles abstraits et leur application aux équations intégrales". In 1922 Banach passed his qualifying examination for the title of docent and in the same year became a professor of the University; two years later he became a corresponding member of the Academy of Learning.

As a professor of the University in Lwów, Banach did important research work. In a short time he became the greatest authority in functional analysis, of which he was one of the creators. Under his and Steinhaus's direction they developed the new Lwów School of Mathematics, which in a short time published its own journal devoted to functional analysis, "Studia Mathematica".

In 1932 the most famous of Banach's work appeared, "Théorie des opérations linéaires" in a new publication called "Mathematical Monographs" of which Banach was one of the founders. That work contributed, to a large extent, to the popularization of Banach's achievements and to the development of functional analysis. The fact that Banach was entrusted with one of the plenary lectures at the International Mathematical Congress in Oslo in 1936 testifies to the mathematical world's interest in Banach. Moreover, he was awarded many scientific prizes and in 1939 he was elected President of the Polish Mathematical Society.

He passed the war years in Lwów. In the years 1940-1941 he was the Dean of the University. After the liberation of Lwów by the Soviet Army he maintained strong contact with Soviet mathematicians. Unfortunately, by that time he had already been stricken by a fatal illness, lung cancer. The death of Stefan Banach, the most brilliant Polish mathematician, took place on 31 August.

A translation in $\mathbb{R}^n$ is a simple example showing that the Banach theorem does not hold if we relax the condition $k < 1$ letting k to be equal to 1; that is, $d(Tx, Ty) \leq d(x, y)$ for every distinct pair of points x, y in X. Such mappings are called *nonexpansive*. Even the "middle" condition $d(Tx, Ty) < d(x, y)$ for every x, y in X (in this case T is usually called *weakly contractive*) does not assure the existence of a fixed point (consider, for instance, the mapping $Tx = x + 1/x$ defined in the complete metric space $[1, +\infty)$ with the euclidean norm). In this situation, it is not surprising that for almost forty years the problem of the existence of a fixed point for nonexpansive mappings was neglected. However, in 1965, Browder [Br1] proved that every nonexpansive mapping T from a convex bounded closed subset C of a Hilbert space X into C has a fixed point. In the same year Browder [Br2], Göhde [Go] and Kirk [Ki1] proved that this result could be improved assuming the weaker condition X is a uniformly convex space or X is a reflexive Banach space with normal structure. These results are noteworthy for the conditions imposed on C which look more suitable for the topological fixed point theory (Schauder's Theorem) and for the "geometric" conditions which X is required to satisfy. From this starting point a very wide theory has been developed which tries to find more general conditions on the Banach space X and the subset C which still assure the existence of fixed points. For simplicity, we shall say that a Banach space X has the *fixed point property* (f.p.p.) if every nonexpansive mapping T defined from a nonempty convex bounded closed subset C of X into C has a fixed point. Since Kakutani [K] showed a simple example of a nonexpansive mapping from the unit ball $\overline{B}(0,1)$ of c_0 into $\overline{B}(0,1)$ without fixed points, it is clear that Banach spaces exist which do not have the f.p.p. The failure of the f.p.p. in this example is a consequence of the noncompactness of $\overline{B}(0,1)$ in the weak topology. However, we shall prove in Chapter VII that every nonexpansive mapping from a weakly compact convex set C of c_0 into C has a fixed point. When such a condition is satisfied we shall say that the Banach space X has the *weak fixed point property* (w.f.p.p.). Obviously, the f.p.p. and the w.f.p.p. are identical if X is reflexive. For a long time an open question was: Does every Banach space X have the w.f.p.p.? The answer to this question was given by Alspach [Al] in 1981, proving that $L^1[0,1]$ fails to have the w.f.p.p. Since every reflexive subspace of $L^1[0,1]$ has the f.p.p. [Mu], one question becomes very important: Does every reflexive Banach space have the f.p.p.? Until now nobody has been able to answer this question.

In Section 1 of this chapter we shall prove that every Banach space with weak normal structure has the w.f.p.p. In Sections 2, 3 and 4 we define the normal structure coefficient and we study bounds from below for this coefficient or certain other similar coefficient which are consequences of various geometric properties of the Banach spaces. In this way we shall find wide classes of spaces with the f.p.p. or w.f.p.p. In particular we shall demonstrate the role of the measures of noncompactness (specifically the modulus of NUC) in metric fixed point theory. In Section 5 we shall study the permanence properties of normal structure under (infinite) direct sum and in Section 6 we compute the normal structure coefficients of the L^p-spaces.

1. Existence of fixed points for nonexpansive mappings: Kirk's Theorem

We start by recalling some definitions.

DEFINITION 1.1. *Let (X,d) be a metric space. A mapping $T : X \to X$ is called nonexpansive if*
$$d(Tx, Ty) \leq d(x, y)$$
for every $x, y \in X$.

DEFINITION 1.2. *Let X be a Banach space, A a bounded subset of X and B an arbitrary subset of X. The Chebyshev radius of set A with respect to set B is* *defined by*

$$r(A, B) = \inf\{\sup\{\|x - y\| : x \in A\} : y \in B\}$$

and we write $r(A)$ for $r(A, \; co(A))$. The Chebyshev centre of A with respect to B is defined by

$$Z(A, B) = \{y \in B : \sup\{\|x - y\| : x \in A\} = r(A, B)\}$$

and we denote $Z(A, \; co(A))$ by $Z(A)$.

Remark 1.3. The set $Z(A, B)$ can be empty. However if for $\varepsilon > 0$ we let

$$Z_\varepsilon(A, B) = \{y \in B : r(A, y) \le r(A, B) + \varepsilon\},$$

then $Z_\varepsilon(A, B)$ is a convex, closed and nonempty set if B is closed nonempty and convex. Thus $Z_\varepsilon(A, B)$ is a convex and weakly compact set if B satisfies the same condition. Since

$$\bigcap_{\varepsilon > 0} Z_\varepsilon(A, B) = Z(A, B),$$

the finite intersection property assures that the Chebyshev centre $Z(A, B)$ is non-empty when B is a convex weakly compact set.

DEFINITION 1.4. *A convex closed bounded set A of a Banach space X is called diametral if $diam(A) = r(A)$. Equivalently, if $Z(A) = A$. We say that the Banach space X has normal structure (respectively weak normal structure) if every closed nonempty bounded (respectively weakly compact) convex diametral subset of X is a singleton.*

Example 1: We shall show throughout this chapter that ℓ^p and $L^p(\Omega)$, $1 < p < +\infty$, have normal structure. The space c_0 does not have normal structure. Indeed, consider the set $A = \overline{co}(\{e_n : n \in \mathbb{N}\})$ where $\{e_n\}$ is the standard basis. Then $diam(A) = 1$ and $r(A) = 1$ because $\lim_{n \to \infty} \|x - e_n\| \ge 1$ for every $x \in c_0$. Since the sequence $\{e_n\}$ is weakly null, A is a weakly compact set and so c_0 does not have weak normal structure either. Considering the same set A in ℓ^1, it is easy to check that ℓ^1 does not have normal structure. However we shall prove in Theorem 3.3 that ℓ^1 (and every Banach space with the Schur property) has weak normal structure.

THEOREM 1.5. *Let X be a Banach space with weak normal structure, C a weakly compact convex subset of X and $T : C \to C$ a nonexpansive mapping. Then T has a fixed point.*

Proof. Let $\mathcal{B}$ be the family of all weakly compact convex nonempty subsets of C which are invariant under T. If we order these sets by inclusion, it is easy to check that this family is inductive. Hence by Zorn's lemma there is a minimal set K in this family. Since $T(K) \subset K$ we have $\overline{co}(T(K)) \subset K$. Thus $\overline{co}(T(K))$ is a weakly compact convex subset of K which is also invariant under T. The minimality of K implies $K = \overline{co}(T(K))$. Since K is a weakly compact and convex set the considerations in Remark 1.3 imply that $Z(K)$ is a nonempty set. Let $x \in Z(K)$, that is, $r(K,x) = r(K)$. For every $y \in K$ we have $\|Ty - Tx\| \leq \|y - x\| \leq r(K)$. Thus $T(K)$ is contained in the closed ball $\overline{B}(Tx, r(K))$ which implies that $\overline{co}(T(K)) = K \subset \overline{B}(Tx, r(K))$. Hence $r(K, Tx) \leq r(K)$ which means $Tx \in Z(K)$. Thus $Z(K)$ is a convex weakly compact subset of K and is invariant under T. Again the minimality of K implies $Z(K) = K$. The weak normal structure of X implies that $\mathrm{diam}(K) = 0$ and so K consists of a fixed point of T. $\square$

We shall include an application of Theorem 1.5 to prove the existence of periodic solutions for a differential equation.

THEOREM 1.6. *Let H be a Hilbert space, $f : [0, +\infty) \times H \to H$ a function which is ω-periodic on the first variable and satisfies the following additional conditions:*
   (i) $f \in C^1([0, +\infty) \times H)$ and is bounded
   (ii) Monotonicity, that is,

$$(f(t, x) - f(t, y)) \cdot (x - y) \leq 0$$

for every $t \in [0, +\infty)$, $x, y \in H$.
   (iii) Inward on the boundary of $\overline{B}(0, 1)$, that is,

$$f(t, x) \cdot x < 0$$

for every $t \in [0, +\infty)$ and $x \in H$ with $\|x\| = 1$.
   Then the differential equation

$$x'(t) = f(t, x(t)) \tag{1}$$

has an ω-periodic solution.

Proof. Since the classic Picard-Lindelöf Theorem does hold in infinite-dimensional spaces [Sc, Theorem V.6.1, page 746], condition (i) assures that (1) has a unique local solution for any initial value problem $x(t_0) = x_0$, $t_0 \in [0, +\infty)$, $x_0 \in \overline{B}(0, 1)$. We can follow an argument as that in the proof of Theorem I.1.11 to prove that any solution can be extended to $[0, +\infty)$. Indeed, if $x(t)$ is a maximal solution in $[0, t_1)$, condition (iii) implies that $x(t) \in \overline{B}(0, 1)$ for every $t \in [0, t_1)$. If $t < t_1$, the boundedness of f lets prove that $\lim_{t \to t_1} x(t) = x_1$ does exist. Since the solution can be extended on a neighborhood of (t_1, x_1) and $x(t)$ is maximal we derive that

$t_1 = +\infty$. We define the Poincaré operator $T : \overline{B}(0,1) \to \overline{B}(0,1)$ by $T(x_0) = x(\omega)$ if $x(t)$ is the solution of (1) which satisfies $x(0) = x_0$. We claim that T is a nonexpansive mapping. Indeed, if $x(t)$ and $y(t)$ are solutions of (1) we have

$$\frac{d}{dt}\left(\|x(t) - y(t)\|\right)^2 = 2(x'(t) - y'(t)) \cdot (x(t) - y(t))$$

$$= 2(f(t, x(t)) - f(t, y(t))) \cdot (x(t) - \dot{y}(t)) \le 0.$$

From this inequality we easily obtain $\|x(t) - y(t)\| \le \|x(0) - y(0)\|$. Thus T is nonexpansive. Since every Hilbert space has normal structure (see, for instance, Theorem 2.2), Theorem 1.5 assures that T has a fixed point which is an ω-periodic solution of (1). □

2. The coefficient $N(X)$ and its connection with uniform convexity

One approach for establishing that a Banach space satisfies the fixed point property (f.p.p.) may be to prove that the space is near (in the sense of Banach-Mazur distance) to another Banach space with the f.p.p. This method needs the use of a "measure" in the sense: To what degree does a Banach space have the f.p.p.? In this way a technique was initiated by Bynum [By3] with the definition of certain normal structure coefficients.

DEFINITION 2.1. *Let X be a Banach space. The most simple normal structure coefficient is the following:*

$$N(X) = \inf\left\{\frac{diam(A)}{r(A)} : A \subset X \quad convex\ closed\ and\ bounded\ with\ \ diam(A) > 0\right\}.$$

It is obvious from the definition that X has normal structure if $N(X) > 1$. We shall later see (Example 5) that the converse result does not hold. We shall say ([E, Ma]) that X has *uniform normal structure* if $N(X) > 1$.

We start by showing that every uniformly convex Banach space has uniform normal structure and $N(X)$ can be bounded from below using the Clarkson modulus.

THEOREM 2.2. *If X is a Banach space with modulus of convexity δ_X, then $N(X) \ge (1 - \delta_X(1))^{-1}$.*

Proof. Let A be a closed convex bounded subset of X with more than one member, and let $\varepsilon > 0$. We denote $d = diam(A)$ and $r = r(A)$. We choose x and y in A such that $\|x - y\| \ge d - \varepsilon$. Let $w = (x+y)/2$, and choose z in A such that $\|z - w\| \ge r - \varepsilon$. Since $\|(z - x)/d\| \le 1$, $\|(z - y)/d\| \le 1$ and $\|(z - x)/d - (z - y)/d\| > (d - \varepsilon)/d$ we obtain through the definition of δ_X that

$$\|z - w\| \le d\left(1 - \delta_X\left(\frac{d - \varepsilon}{d}\right)\right).$$

Thus

$$r \leq \varepsilon + d \left(1 - \delta_X \left(\frac{d - \varepsilon}{d} \right) \right)$$

which from the continuity of δ_X implies the desired result.   □

Remark 2.3.

(a) Let X be a Banach space. For each $z \in X, \|z\| = 1$ the following convexity modulus can be considered: $\delta_z(\varepsilon) = \inf\{1 - \|x + y\|/2 : \|x\| \leq 1, \|y\| \leq 1, \|x - y\| \geq \varepsilon, x - y = tz$ for some real $t\}$. If $\delta_z(\varepsilon) > 0$ for all $\varepsilon > 0$ and all z then X is called *uniformly convex in every direction* (UCED). This notion was introduced by Garkavi [Gar] to characterize those Banach spaces such that $Z(K)$ is a singleton for every bounded subset K of X. In [DJS] it is proved that UCED spaces have normal structure. Indeed, assume that A is a bounded convex subset of a UCED space X with more than one member and write $\text{diam}(A) = d$. If ε is an arbitrary positive number we can choose $x, y \in A$ such that $\|x - y\| \geq d - \varepsilon$. Let $z = (x - y)/\|x - y\|$. Then for every $w \in A$ we have

$$\|w - (x + y)/2\| \leq d \left(1 - \delta_z \left(\frac{d - \varepsilon}{d} \right) \right) < d.$$

Thus $(x + y)/2$ is not a diametral point and X has normal structure. However $N(X)$ can be 1 for a UCED space (see Example 5).

(b) The bound which is given in Theorem 2.2 is not very sharp in certain Banach spaces. For instance, $N(\ell^2) = \sqrt{2}$ (see Theorem 6.3). However in this space the Clarkson modulus is $\delta_{\ell^2}(\varepsilon) = 1 - \sqrt{1 - \varepsilon^2/4}$ which gives the bound $2/\sqrt{3}$.

(c) Theorem 2.2 proves that X has normal structure when $\varepsilon_0(X) < 1$. What is the situation when $\varepsilon_0(X) \geq 1$? The following example shows that all situations are possible.

Example 2: Let $\ell^{p,q}$ be the Banach spaces introduced in Example II.3. We know (Example IV.7) that $\varepsilon_0(\ell^{p,1}) = 2^{1/p}$. We shall prove in Theorem 3.11 that $\ell^{p,1}$ has normal structure. This property can also be deduced keeping in mind that $\ell^{p,1}$ is 2-UC (Example IV.8) and by applying Theorem 2.5, where it is proved that every k-UC space has uniform normal structure. On the other hand, the space $\ell^{p,\infty}$ fails to have normal structure. Indeed the basic sequence is a diametral sequence because $\|e_n - e_m\|_{p,\infty} = 1$ if $n \neq m$ and for every point $u = \sum_{i=1}^{n} \alpha_i e_i, \alpha_i \geq 0, \sum_{i=1}^{n} \alpha_i = 1$ we have $\|e_{n+1} - u\|_{p,\infty}^p = \sup\{1, \sum \alpha_i^p\} = 1$. However, we proved in Example IV.7 that $\varepsilon_0(\ell^{p,\infty}) = 1$. Furthermore, the spaces $\ell^{p,\infty}$ and $\ell^{p,1}$ are dual to each other. Thus normal structure is not an invariant property under passage to dual spaces.

Following on, we shall prove that every k-uniform convex space has uniform normal structure [Su, Am]. We need an easy lemma:

LEMMA 2.4. *Let ε and η be numbers in $(0, 1)$. Assume that $r \in [0, 1]$ satisfies $r > 1 - (1 - \varepsilon - \eta)/n!n$. Then $(1 - \eta)r^{n-1} - (n! - 1)n(1 - r) > \varepsilon$.*

Proof. Note that if r satisfies the above condition, then

$$(1-\eta)r^{n-1} - (n! - 1)n(1-r) > (1-\eta)r^{n-1} - 1 + \varepsilon + \eta + n(1-r).$$

Thus it suffices to prove $(1-\eta)r^{n-1} - 1 + \eta + n(1-r) \geq 0$ for every $r \in [0,1]$. Since the result is obvious if $n = 1$ we can assume $n > 1$. Denote $f(r) = (1-\eta)r^{n-1} - 1 + \eta + n(1-r)$. Then $f(0) > 0$ and $f(1) = 0$. Since $f'(r) \neq 0$ in $[0,1]$ we obtain $f(r) \geq 0$ in this interval. $\qquad\square$

THEOREM 2.5. *Let X be a Banach space. Then*

$$N(X) \geq \frac{1}{\max\left\{1 - \frac{1-\varepsilon}{k!k}, 1 - \delta_X^k(\varepsilon)\right\}}.$$

In particular, X has uniform normal structure if $\varepsilon_0^k(X) < 1$.

Proof. Let $A \subset X$ be a convex set with $\mathrm{diam}(A) = 1$. We choose any $r < r(A)$ and any positive η. We can find $x_0, x_1 \in A$ such that $\|x_1 - x_0\| > 1 - \eta$. Since $r < r(A)$ for every $j = 2, 3, \ldots, k+1$, $x_j \in A$ can be chosen such that $\|x_j - j^{-1} \sum_{i=0}^{j-1} x_i\| > r$. By translation we can assume $x_{k+1} = 0$. So every x_j, $j = 0, 1, \ldots, k$ lies in $\overline{B}(0,1)$. We can choose norm one functionals $f_1, \ldots, f_k$ in X^* such that $f_1(x_1 - x_0) > 1 - \eta$ and $f_j\left(x_j - j^{-1} \sum_{i=0}^{j-1} x_i\right) > r$. Now, we can bound the n-dimensional volume of $\mathrm{co}(\{x_0, \ldots, x_k\})$. Indeed

$$A(x_0,\ldots,x_k) \geq \det \begin{vmatrix} 1 & \cdots & 1 \\ f_1(x_0) & \cdots & f_1(x_k) \\ \vdots & \vdots & \vdots \\ f_k(x_0) & \cdots & f_k(x_k) \end{vmatrix}$$

$$= \det \begin{vmatrix} f_1(x_1-x_0) & f_1(x_2-\frac{1}{2}(x_0+x_1)) & \cdots & f_1(x_k-\frac{1}{k}\sum_{i=0}^{k-1} x_i) \\ \vdots & \vdots & \vdots \\ f_k(x_1-x_0) & f_k(x_2-\frac{1}{2}(x_0+x_1)) & \cdots & f_k(x_k-\frac{1}{k}\sum_{i=0}^{k-1} x_i) \end{vmatrix}.$$

All the entries in this determinant have absolute value ≤ 1. Furthermore, since $m^{-1} \sum_{i=0}^{m-1} f_m(x_m - x_i) > r$ and $|f_m(x_m - x_i)| \leq 1$, we have $1 - m(1-r) < f_m(x_m - x_i) \leq 1$ for $i < m$. Hence

$$|f_m(x_j - x_i)| = |f_m(x_m - x_i) - f_m(x_m - x_j)| < m(1-r)$$

and

$$\left| f_m\left(x_j - \frac{1}{j}\sum_{i=0}^{j-1} x_i\right)\right| = \left|\frac{1}{j}\sum_{1=0}^{j-1} f_m(x_j - x_i)\right| < m(1-r).$$

Thus

$$A(x_0,\ldots,x_k) > (1-\eta)r^{k-1} - (k! - 1)k(1-r)$$

and using Lemma 2.4, $(1-\eta)r^{k-1} - (k!-1)k(1-r) < \varepsilon$ if $r > 1 - (1-\varepsilon-\eta)/k!k$. Therefore for such r we have

$$r < \left\| \frac{1}{k+1} \sum_{i=0}^{k} x_i \right\| < 1 - \delta_X^k(\varepsilon).$$

Since $\eta > 0$ and $r < r(A)$ are arbitrary, we obtain $r(A) \leq \max\{1 - \delta_X^k(\varepsilon), 1 - (1-\varepsilon)/k!k\}$. □

It is worth noting that when $N(X) > 1$ the space X is reflexive.

THEOREM 2.6. *Let X be a Banach space such that $N(X) > 1$. Then X is reflexive.*

Proof. If X is not reflexive, for every $\varepsilon > 0$ there exists a sequence $\{x_n\}$ (see [MM]) such that $1 - \varepsilon \leq \|u_{1,n} - u_{n,\omega}\| \leq 1 + \varepsilon$ for any $u_{1,n} \in \text{co}(\{x_j\}_{1\leq j\leq n})$, $u_{n,\omega} \in \text{co}(\{x_j\}_{j>n})$, and for any n. Then, it is clear that $\text{diam}(\{x_n\}) \leq 1 + \varepsilon$. On the other hand if v belongs to $\text{co}(\{x_n\})$ and n is large enough we have $\|x_n - v\| \geq 1 - \varepsilon$. Since ε is arbitrary we obtain $N(X) = 1$. □

Remark 2.7. Since $N(X) > 1$ if $\delta_X^k(1) > 0$, Theorems 2.5 and 2.6 improve Theorem IV.2.19. The converse of Theorem 2.6 does not hold. Indeed, Example 6 shows a reflexive Banach space X with $N(X) = 1$.

The following result shows how the normal structure coefficients can be useful for proving the stability of the fixed point property. We recall that for two isomorphic Banach spaces X and Y, *Banach-Mazur distance* is defined by

$$d(X,Y) = \inf \left\{ \|T\|\|T^{-1}\| : T \in \text{Isom}(X,Y) \right\}.$$

It is clear that $d(X,Y) = 1$ when X and Y are isometric.

THEOREM 2.8. *Let X and Y be isomorphic Banach spaces, then*

$$N(X) \leq d(X,Y)N(Y).$$

Proof. Let C be a closed bounded convex subset of Y. If $U : Y \to X$ is an isomorphism we have

$$r(C) \leq \|U^{-1}\| r(U(C)) \leq \|U^{-1}\| \, \text{diam}(U(C))/N(X) \leq \|U^{-1}\|\|U\| \, \text{diam}(C)/N(X).$$

Thus $r(C) \leq d(X,Y) \, \text{diam}(C)/N(X)$ which implies the desired result. □

3. The weakly convergent sequence coefficient

We shall now consider another normal structure coefficient. Before introducing this coefficient we shall detail the notions and notations which will be used.

DEFINITION 3.1. *The asymptotic diameter, radius and centre of a sequence $\{x_n\}$ in a Banach space X will be defined by:*

$$diam_a(\{x_n\}) = \lim_{k \to \infty} \sup\{\|x_n - x_m\| : n, m \geq k\},$$

$$r_a(\{x_n\}, B) = \inf\{\limsup_{n \to \infty} \|x_n - y\| : y \in B\},$$

$$Z_a(\{x_n\}, B) = \{y \in B : \limsup_{n \to \infty} \|x_n - y\| = r_a(\{x_n\}, B)\},$$

for a subset B of X. When $B = \overline{co}(\{x_n\})$ we will denote $r_a(\{x_n\}, \overline{co}(\{x_n\}))$ and $Z_a(\{x_n\}, \overline{co}(\{x_n\}))$ respectively by $r_a(\{x_n\})$ and $Z_a(\{x_n\})$.

DEFINITION 3.2. *Let X be a Banach space without the Schur property, that is, there exist weakly convergent sequences which are not norm convergent. The weakly convergent sequence coefficient of a Banach space X is defined by*

$$WCS(X) = \inf\{\frac{diam_a(\{x_n\})}{r_a(\{x_n\})} : \{x_n\} \text{ is a weakly convergent sequence}$$

$$\text{which is not norm convergent}\}$$

Since 2 is the maximum value for $WCS(X)$ in Definition 3.1 we shall say that $WCS(X) = 2$ when X satisfies the Schur property. We shall prove that the weakly convergent sequence coefficient can be considered as a measure of the weak normal structure of X.

THEOREM 3.3. *Let X be a Banach space with $WCS(X) > 1$. Then X has weak normal structure, that is, every weakly compact convex subset of X with more than one member is not diametral.*

Proof. Assume that X contains a diametral weakly compact convex set A with more than one member. Denote $d = \text{diam}(A) > 0$ and let $\varepsilon < d$ be an arbitrary positive number. Choose an arbitrary x_1 in A. By induction we can construct a sequence $\{x_n\}$ such that

$$\|y_n - x_{n+1}\| > d - \frac{\varepsilon}{n^2}$$

where $y_n = \sum_{i=1}^{n} x_i/n$. Let x be an arbitrary point in the convex hull of $\{x_1, \dots, x_n\}$, that is, $x = \sum_{j=1}^{n} \alpha_j x_j$ where $\alpha_j \geq 0$ and $\sum_{j=1}^{n} \alpha_j = 1$. If $\alpha = \alpha_p = \max\{\alpha_1, \dots, \alpha_n\}$, then

$$y_n = \frac{x}{n\alpha} + \sum_{j=1}^{n} \left(\frac{1}{n} - \frac{\alpha_j}{n\alpha}\right) x_j.$$

Since

$$\frac{1}{n\alpha} + \sum_{j=1}^{n} \left(\frac{1}{n} - \frac{\alpha_j}{n\alpha}\right) = 1 \quad ; \quad \frac{1}{n} - \frac{\alpha_j}{n\alpha} \geq 0$$

we have

$$d - \frac{\varepsilon}{n^2} < \|y_n - x_{n+1}\| \leq \frac{1}{n\alpha}\|x - x_{n+1}\| + \sum_{j=1}^{n} \left(\frac{1}{n} - \frac{\alpha_j}{n\alpha}\right)\|x_j - x_{n+1}\|$$

$$\leq \frac{1}{n\alpha}\|x - x_{n+1}\| + \left(1 - \frac{1}{n\alpha}\right)d.$$

Hence

$$\|x - x_{n+1}\| \geq \left(\frac{d}{n\alpha} - \frac{\varepsilon}{n^2}\right)n\alpha = d - \frac{\varepsilon\alpha}{n} \geq d - \frac{\varepsilon}{n}.$$

Thus $\lim_{n\to\infty} d(x_{n+1}, \mathrm{co}(\{x_1, \ldots, x_n\})) = d$. Since A is weakly compact and every subsequence of $\{x_n\}$ satisfies the same condition, we can assume that $\{x_n\}$ is weakly convergent. In particular $\mathrm{diam}_a(\{x_n\}) \leq d$. If X satisfies the Schur property, $\{x_n\}$ is convergent and we obtain the contradiction $d = 0$. Otherwise, if y belongs to the convex hull of $\{x_n\}$ we know that y belongs to $\mathrm{co}(\{x_1, \ldots, x_k\})$ for certain k. For $n > k$ we have $\|y - x_n\| \geq d - \varepsilon/n$. Hence $r_a(\{x_n\}) \geq d$. Since $\mathrm{diam}_a(\{x_n\}) \leq d$ we obtain $WCS(X) \leq 1$. $\qquad\square$

Remark 3.4. Example 5 will show that $WCS(X)$ can be equal to one for a space with weak normal structure. We shall say that X has *uniform weak normal structure* if $WCS(X) > 1$. According to Theorem 3.3 reflexive Banach spaces with uniform weak normal structure have normal structure. However, we shall show in Example 6 a reflexive Banach space with uniform weak normal structure which does not have uniform normal structure.

The following result shows that $WCS(X)$ can be useful for studying the stability of the fixed point property. It can be proved in a similar way as Theorem 2.8.

THEOREM 3.5. *Let X and Y be isomorphic Banach spaces, then*

$$WCS(X) \leq d(X, Y)WCS(Y).$$

Remark 3.6. According to the above theorem, a reflexive Banach space Y has the f.p.p. if $d(X, Y) < WCS(X)$ for some Banach space X. Some improvements of this result can be found in [By3] and [Pr1].

We are going to state a more convenient form for $WCS(X)$ in reflexive spaces. Let T be a topological space and $f : T \to \mathbb{R}$. We recall that f is *lower semicontinuous* at $a \in T$ if for every $\varepsilon > 0$ the set $f^{-1}((f(a) - \varepsilon, +\infty))$ is a neighbourhood of a. In this case we have $\liminf_{n\to\infty} f(a_n) \geq f(a)$ if $\{a_n\}$ is a sequence convergent to a. Thus, if T is a compact topological space and f is lower semicontinuous in T, f attains an absolute minimum at a point of T.

LEMMA 3.7. *Let $\{x_n\}$ be a bounded sequence in a Banach space X. Define $\Phi :$ $X \to \mathbb{R}$ by $\Phi(x) = \limsup_{n\to\infty} \|x_n - x\|$. Then Φ is a weakly lower semicontinuous function.*

Proof. It is clear that Φ is (strongly) continuous. Then $\Phi^{-1}((-\infty, a])$ is a closed set for every $a \in \mathbb{R}$. Since Φ is a convex function it is clear that $\Phi^{-1}((-\infty, a])$ is also a convex set. Thus, it is a weakly closed set and $\Phi^{-1}((a, +\infty))$ is a weakly open set. $\qquad\square$

LEMMA 3.8. *Let X be a Banach space without the Schur property. Then:*

(a)

$$WCS(X) = \inf \left\{ \frac{diam_a(\{x_n\})}{\limsup_{n\to\infty} \|x_n\|} : \{x_n\} \text{ converges weakly to zero} \right\}.$$

(b)

$$WCS(X) = \inf \{ \frac{\lim_{n,m ; n \neq m} \|x_n - x_m\|}{\lim_{n\to\infty} \|x_n\|} : \{x_n\} \text{ converges weakly to zero and}$$

$$\lim_{n,m ; n \neq m} \|x_n - x_m\| \text{ and } \lim_{n\to\infty} \|x_n\| \text{ exist}\}.$$

(c)

$$WCS(X) = \inf \{ \lim_{n,m ; n \neq m} \|x_n - x_m\| : \{x_n\} \text{ converges weakly to zero,}$$

$$\|x_n\| = 1 \text{ and } \lim_{n,m ; n \neq m} \|x_n - x_m\| \text{ exists}\}.$$

(d)

$$WCS(X) = \inf \{ \frac{a}{\limsup_{n\to+\infty} \|x_n\|} : \{x_n\} \text{ converges weakly}$$

$$\text{to zero }, \lim_{n,m ; n \neq m} \|x_n - x_m\| = a\}.$$

Proof. Let $\{x_n\}$ be a weakly null sequence. For each $k \geq 1$, A_k denotes the closed convex hull of $\{x_n\}_{n\geq k}$. From Lemma V.1.9 we know that $\bigcap_{k=1}^{\infty} A_k = \{0\}$. Since the function $\Phi(x) = \limsup_{n\to\infty} \|x_n - x\|$ is weak lower semicontinuous and A_k is weakly compact, this mapping attains a minimum on A_k. Thus the Chebyshev centre $Z(\{x_n\}, A_k)$ is nonempty. We choose $z_k \in Z(\{x_n\}, A_k)$. Since $\{z_k\}$ is contained in a weakly compact set and 0 is the unique point which can be adherent to $\{z_k\}$, we infer that $\{z_k\}$ is weakly null. Furthermore it is clear that $\{\Phi(z_k)\}$ is a nondecreasing sequence which is bounded by $\Phi(0)$. Thus $\lim_{k\to\infty} \Phi(z_k) \leq \Phi(0)$. On the other hand the weakly lower semicontinuity of Φ implies $\lim_{k\to\infty} \Phi(z_k) \geq \Phi(0)$ and so $\lim_{k\to\infty} \Phi(z_k) = \Phi(0)$. Since

$$\Phi(z_k) = \min_{z\in A_k} \limsup_{n\to\infty} \|x_n - z\| = r_a(\{x_n\}_{n\geq k})$$

we obtain

$$\Phi(z_k)WCS(X) \leq \mathrm{diam}_a(\{x_n\}).$$

Taking the limit in k we obtain

$$\limsup_{n\to\infty} \|x_n\|WCS(X) \leq \mathrm{diam}_a(\{x_n\})$$

and from this inequality we obtain (a). Statements (b), (c) and (d) are now clear because for every sequence we can obtain a subsequence $\{x_n\}$ such that $\lim_{n,m\,;n\neq m} \|x_n - x_m\|$ exists (see Theorem III.1.5). ☐

Lemma 3.8 lets us easily compute the coefficient $WCS(X)$ in ℓ^p-spaces.

THEOREM 3.9. (a) *For every real number $p \geq 1$ we have $WCS(\ell^p) = 2^{1/p}$.*
  (b) $WCS(c_0) = 1$.

Proof. (a) Let $\{x_n\}$ be a weakly null sequence in ℓ^p, $p > 1$, such that $\|x_n\|_p = 1$ for every n and $\lim_{n,m\,;n\neq m} \|x_n - x_m\|_p$ exists. Let ε be an arbitrary positive number. Using standard arguments it is easy to find a subsequence $\{y_n\}$ of $\{x_n\}$ with nearly disjoint supports, that is, an increasing sequence $\{k_n\}$ of positive integers exists such that

$$\sum_{k\geq k_{n+1}} |y_n^k|_p^p + \sum_{k<k_n} |y_n^k|_p^p < \varepsilon.$$

Therefore $\|y_n - y_m\|_p^p \geq \|y_n\|_p^p + \|y_m\|_p^p - o(1)$ where $o(1) \to 0$ as $\varepsilon \to 0$. This implies $\lim_{n,m\,;n\neq m} \|x_n - x_m\|_p \geq (2 - o(1))^{1/p}$. Since ε is arbitrary we obtain $WCS(\ell^p) \geq 2^{1/p}$. Since the basic sequence $\{e_n\}$ satisfies $\mathrm{diam}_a(\{e_n\}) = 2^{1/p}$, $e_n \rightharpoonup 0$ and $\|e_n\|_p = 1$ for every $n \in \mathbb{N}$, we obtain the desired result.
  (b) It is a consequence of Theorem 3.3 and Example 1. ☐

In Example 2 we considered the spaces $\ell^{p,q}$. We shall now compute the weakly convergent sequence coefficient of these spaces.

LEMMA 3.10. (a) *Let a, b, p, q be nonnegative numbers such that $1 \leq q < p$. Then*

$$(a^q + b^q) \leq 2^{(p-q)/p}(a^p + b^p)^{q/p}.$$

  (b) *Let $\{a_1, \ldots, a_n, b_1, \ldots, b_n\}$ be nonnegative numbers, $1 < p < q$. Then*

$$\left(\sum_{n=1}^k a_n^p\right)^{q/p} + \left(\sum_{n=1}^k b_n^p\right)^{q/p} \leq \left(\sum_{n=1}^k (a_n + b_n)^p\right)^{q/p}.$$

Proof. (a) The convexity of the function $x \to x^p$ implies $(x+y)^p \leq 2^{p-1}(x^p + y^p)$. Replacing p by p/q, x by a^q and y by b^q we get the inequality.

(b) Since $q/p > 1$ we have

$$a^{q/p} + b^{q/p} \leq (a+b)^{q/p}$$

for any positive number a and b. On the other hand $(a_n + b_n)^p \geq a_n^p + b_n^p$. Taking $a = \sum_{n=1}^{k} a_n^p$ and $b = \sum_{n=1}^{k} b_n^p$ these inequalities give:

$$\left(\sum_{n=1}^{k} a_n^p \right)^{q/p} + \left(\sum_{n=1}^{k} b_n^p \right)^{q/p} \leq \left(\sum_{n=1}^{k} a_n^p + \sum_{n=1}^{k} b_n^p \right)^{q/p} \leq \left(\sum_{n=1}^{k} (a_n + b_n)^p \right)^{q/p}.$$

$\square$

THEOREM 3.11. *Let p, q be real numbers, $1 < p$, $1 \leq q$. Then $WCS(\ell^{p,q}) = \min\{2^{1/p}, 2^{1/q}\}$.*

Proof. Considering the sequence $\{e_{2n} - e_{2n+1}\}$ in the case $q \leq p$, and the sequence $\{e_n\}$ in the case $q > p$, we see that $WCS(\ell^{p,q}) \leq \min\{2^{1/p}, 2^{1/q}\}$. On the other hand, let $\{u_n\}$ be a weakly null sequence in $\ell^{p,q}$ such that $\lim_{n \to \infty} \|u_n\|_{p,q} = 1$. By a standard method we can construct a sequence $\{x_n\}$ and a subsequence of $\{u_n\}$ such that $\mathrm{supp}(x_n) \cap \mathrm{supp}(x_m) = \emptyset$ if $n \neq m$ and $\lim_{n \to \infty} \|x_n - u_{k_n}\|_{p,q} = 0$. If $q \leq p$, using Lemma 3.10 (a) we obtain

$$\begin{aligned}
\|x_n\|_{p,q}^q + \|x_m\|_{p,q}^q &= (\|x_n^+\|_p^q + \|x_n^-\|_p^q) + (\|x_m^+\|_p^q + \|x_m^-\|_p^q) \\
&= (\|x_n^+\|_p^q + \|x_m^+\|_p^q) + (\|x_n^-\|_p^q + \|x_m^-\|_p^q) \\
&\leq 2^{\frac{p-q}{p}} ((\|x_n^+\|_p^p + \|x_m^+\|_p^p)^{\frac{q}{p}} + (\|x_n^-\|_p^p + \|x_m^-\|_p^p)^{\frac{q}{p}}) \\
&= 2^{\frac{p-q}{p}} (\|x_n^+ + x_m^+\|_p^q + \|x_n^- + x_m^-\|_p^q) \\
&= 2^{\frac{p-q}{p}} \|x_n + x_m\|_{p,q}^q = 2^{\frac{p-q}{p}} \|x_n - x_m\|_{p,q}^q.
\end{aligned}$$

Therefore $2 \leq 2^{(p-q)/p} (\mathrm{diam}_a(\{x_n\}))^q$ which implies $WCS(\ell^{p,q}) \geq 2^{1/p}$.

We now consider $q \geq p$. We only need to prove that

$$\|x_n\|_{p,q}^q + \|x_m\|_{p,q}^q \leq \|x_n - x_m\|_{p,q}^q.$$

From Lemma 3.10 (b) we easily deduce

$$\|x_n^+\|_p^q + \|x_m^+\|_p^q \leq \|x_n^+ + x_m^+\|_p^q \quad ; \quad \|x_n^-\|_p^q + \|x_m^-\|_p^q \leq \|x_n^- + x_m^-\|_p^q$$

which proves the desired result. $\square$

4. Uniform smoothness, near uniform convexity and normal structure

We have seen that $N(X) > 1$ if the characteristic of convexity of X is less than 1. A similar result for $WCS(X)$ will be proved concerning the modulus of uniform smoothness.

THEOREM 4.1. *Let X be a Banach space with modulus of smoothness ρ_X. Denote*

$$\rho = \inf\left\{\rho_X(\tau) - \frac{\tau}{2} + 1 : \tau \in (0, 1/2]\right\}.$$

Then $WCS(X) \geq 1/\rho$. In particular X has normal structure if $\rho'_X(0) < 1/2$.

Proof. We begin by proving that the space X is reflexive if $\rho < 1$. Indeed, according to the Lindenstrauss formula (Theorem IV.3.10) we have $\rho_X(\tau) \geq (\tau/2) - \delta_{X^*}(1)$ for every τ. If $\rho < 1$ for some τ we have $1 > \rho_X(\tau) - (\tau/2) + 1 \geq 1 - \delta_{X^*}(1)$, which implies $\delta_{X^*}(1) > 0$. Hence X^* is reflexive and therefore so is X. Since $\rho \leq 1$ and the result is obvious if $\rho = 1$, we can assume that X is a reflexive space. Let τ be a number in $(0, 1/2]$ and $\{x_n\}$ a normalized weakly null sequence in X. Let $d = \lim_{n,m\,;n\neq m} \|x_n - x_m\|$ and consider a sequence $\{x_n^*\}$ of norm one functionals for which $x_n^*(x_n) = 1$. Since X^* is reflexive we can assume that $\{x_n^*\}$ converges weakly to some $x^* \in X^*$. Let η be an arbitrary positive number and choose n large enough so that $|x^*(x_n)| < \eta/2$ and $d - \eta < \|x_n - x_m\| < d + \eta$ for any $m > n$. Then for a sufficiently large $m > n$ we have

$$|(x_m^* - x^*)(x_n)| < \eta/2 \quad \text{and} \quad |x_n^*(x_m)| < \eta.$$

Thus $|x_m^*(x_n)| < \eta$ and if $l = \|x_n - x_m\| \leq 2$ we have

$$
\begin{aligned}
\rho_X(\tau) &\geq \frac{1}{2}\left(\left\|\frac{x_n - x_m}{l} + \tau x_n\right\| + \left\|\frac{x_n - x_m}{l} - \tau x_n\right\|\right) - 1 \\
&\geq \frac{1}{2}\left(x_n^*\left(\left(\frac{1}{l} + \tau\right)x_n - \frac{x_m}{l}\right) + x_m^*\left(\frac{x_m}{l} - \left(\frac{1}{l} - \tau\right)x_n\right)\right) - 1 \\
&\geq \frac{1}{2}\left(\frac{1}{l} + \tau - \frac{\eta}{l} + \frac{1}{l} - \left(\frac{1}{l} - \tau\right)\eta\right) - 1 \geq \frac{1}{d + \eta} + \frac{\tau}{2} - \frac{\eta}{d - \eta} - 1.
\end{aligned}
$$

Since η is arbitrary we obtain

$$\rho_X(\tau) \geq \frac{1}{d} + \frac{\tau}{2} - 1.$$

Therefore $1/d \leq \rho_X(\tau) - \tau/2 + 1$ for every $\tau \in (0, 1/2]$ and so $d \geq 1/\rho$. Thus $WCS(X) \geq 1/\rho$. If $\rho'_X(0) < 1/2$ it is clear that there exists $\tau \in (0, 1/2]$ such that $\rho_X(\tau)/\tau < 1/2$. Hence $\rho < 1$, $WCS(X) > 1$ and so X has normal structure. $\square$

COROLLARY 4.2. *Every US space has normal structure.*

Remark 4.3. A similar estimate for the normal structure coefficient is proved in [Pr4], that is, $N(X) \geq 1/\rho$ for every Banach space X.

We have seen that the Clarkson modulus of convexity gives a lower bound for the normal structure coefficient. The next step will be to prove that a similar bound can be obtained for the weakly convergent sequence coefficient replacing the Clarkson modulus with the modulus of noncompact convexity corresponding to the separation measure of noncompactness.

THEOREM 4.4. *Let X be a Banach space. Then*

$$WCS(X) \geq \lim_{\varepsilon \to 1^-} \frac{1}{1 - \Delta_{X,\beta}(\varepsilon)}.$$

Proof. If X is not reflexive, $\lim_{\varepsilon \to 1^-} \Delta_{X,\beta}(\varepsilon) = 0$ (Theorem V.1.7) and the result is obvious. Thus we can assume that X is reflexive. Let $\varepsilon < 1$ be an arbitrary positive number and $\{x_n\}$ a sequence weakly convergent to 0 such that $\lim_{n,m \,;n \neq m} \|x_n - x_m\| = (\varepsilon+1)/2$. For any k consider the sequence $\{x_k - x_n\}_{n>k}$. We have Sep $(\{x_k - x_n\}_{n>k}) > \varepsilon$ and $\{x_k - x_n\}_{n>k} \subset \overline{B}(0,1)$ for a large enough k. Since $\{x_k - x_n\}_{n>k}$ converges weakly to x_k we obtain $\|x_k\| \leq 1 - \Delta_{X,\beta}(\varepsilon)$. By Lemma 3.8 (d) we obtain

$$WCS(X) \geq \frac{\frac{\varepsilon+1}{2}}{1 - \Delta_{X,\beta}(\varepsilon)}.$$

Taking the limit as $\varepsilon \longmapsto 1^-$ we conclude the proof. $\qquad\square$

Remark 4.5. In the particular case when $X = \ell^p$, from Theorem 4.4 and Theorem IV.1.16 we obtain $WCS(\ell^p) \geq 2^{\frac{1}{p}}$, that is, the actual value of $WCS(\ell^p)$ (see Theorem 3.9). Thus this bound is the best possible in general. Furthermore this bound does not hold for $N(X)$ because in Theorem 6.3 we shall show that $N(\ell^p) = 2^{1-\frac{1}{p}} < 2^{\frac{1}{p}}$ if $1 < p < 2$.

In some spaces, $WCS(X)$ can be strictly greater than the lower bound obtained in Theorem 4.4, as the following example shows.

Example 3: Let X be the space ℓ^2 renormed by

$$\|(x^k)\| = \max\left\{ |x^1|, \left(\sum_{k=2}^{\infty} (x^k)^2 \right)^{\frac{1}{2}} \right\}.$$

If $\{x_n\} = \{(x_n^k)\}$ is a weakly convergent sequence, taking a subsequence and by translation we can assume that $\{x_n^1\} \to 0$. So $WCS(X) = WCS(\ell^2) = \sqrt{2}$. However, considering the sequence $\{x_n\} = \{(x_n^k)\}$ where $x_n^1 = 1$ for every $n \in \mathbb{N}$, and $x_n^k = \delta_{kn}$ for $k \geq 2$, it is clear that $\Delta_{X,\beta}(1) = 0$.

From Theorem 4.4 we can obtain the following result:

COROLLARY 4.7. *Let X be a Banach space and $\varepsilon_\beta(X) < 1$. Then X has normal structure.*

Proof. If $\varepsilon_\beta(X) < 1$ then $\lim_{\varepsilon \to 1^-} (1 - \Delta_{X,\beta}(\varepsilon))^{-1} > 1$. Thus $WCS(X) > 1$ which implies that X has normal structure. $\qquad\square$

Remark 4.7. The above corollary implies that every NUC space has normal structure. On the other hand we have seen that every uniformly smooth space has normal structure. The situation for NUS spaces is different. Indeed, we know that $\ell^{p,1}$ is 2-UC and so it is NUC. Thus its dual $\ell^{q,\infty}$ is NUS but this space fails to have normal structure (Example 2). However it can be proved that NUS spaces also have the fixed point property [Ga] (see Corollary VII.2.10).

5. Normal structure in direct sum spaces

Theorem 3.9 can be considered as a special case of the following theorem, where the weakly convergent sequence coefficient is computed in ℓ^p direct sum of Banach spaces. We need to recall some definitions:

DEFINITION 5.1. *Let $\{X_n, |\cdot|_n\}$ be a sequence of Banach spaces, $1 \le p < \infty$. The Banach space*

$$\bigoplus_p X_n = \left\{ x = (x^n) \in \bigotimes X_n : \sum_{n=1}^\infty |x^n|_n^p < \infty \right\}$$

with the norm $\|x\| = (\sum_{n=1}^\infty |x^n|_n^p)^{1/p}$ will be called the ℓ^p-direct sum of X_n.

For $p > 1$ it is well known that $\bigoplus_p X_n$ is reflexive if every X_n is reflexive and the conjugate space is $\bigoplus_q X_n^*$, where $q = p/(p-1)$.

THEOREM 5.2. *Let $\{X_i\}$ be a sequence of reflexive Banach spaces. Then, for any $p \in (1, +\infty)$ we have*

$$WCS(\bigoplus_p X_i) = \inf\{WCS(X_i), 2^{1/p} : i \in \mathbb{N}\}.$$

Proof. Since every space X_i is a subspace of $\bigoplus_p X_i$, it is clear that $WCS(\bigoplus_p X_i)$ $\le \inf\{WCS(X_i) : i \in \mathbb{N}\}$. On the other hand consider the sequence $\{x_n\}$ in $\bigoplus_p X_i$ where x_n^n is a norm one vector in X_n and $x_n^i = 0$ if $n \ne i$. Then $\{x_n\}$ is weakly null, $\|x_n\| = 1$ for every n and $\|x_n - x_m\| = 2^{1/p}$ if $n \ne m$. Thus $WCS(\bigoplus_p X_i) \le WCS(\ell^p)$. To prove the converse inequality, bearing Lemma 3.8 (c) in mind, we only need to prove that $l \ge \inf\{2^{1/p}, WCS(X_i) : i \in \mathbb{N}\}$ for every normalized weakly null sequence $\{x_n\}$ in $\bigoplus_p X_i$ such that $\lim_{n,m\,;n\ne m} \|x_n - x_m\| = l$. Let us denote by w the $\inf\{2^{1/p}, WCS(X_i) : i \in \mathbb{N}\}$ and let $\{x_n\}$ be a normalized weakly null sequence in $\bigoplus_p X_i$ such that $\lim_{n,m\,;n\neq m} \|x_n - x_m\| = l$. Taking sub-sequences and using a diagonal argument we can assume that $\lim_{n\to\infty} |x_n^i|_i = a^i$ and $\lim_{n,m;n\neq m} \|x_n^i - x_m^i\|_i = l_i$ for any $i \in \mathbb{N}$. Since $\sum_{i=1}^{\infty} |x_n^i|_i^p = 1$ it is easy to check that (a^i) is in ℓ^p and $\|(a^i)\|_p \leq 1$. Let ε be a positive number. There exists $i_1 \in \mathbb{N}$ such that $\sum_{i>i_1}(a^i)^p < \varepsilon$. Choose a large enough $n_1 \in \mathbb{N}$ such that $\sum_{i\leq i_1}\||x_m^i|_i - a^i|^p < \varepsilon$, $\|x_{n_1} - x_m\|^p < l^p + \varepsilon$ and $\sum_{i\leq i_1}\||x_{n_1}^i - x_m^i|_i - l_i|^p < \varepsilon$ for any $m > n_1$. There exists $i_2 \in \mathbb{N}$ such that $\sum_{i>i_2}|x_{n_1}^i|_i^p < \varepsilon$. We choose a large enough $n_2 \in \mathbb{N}$ such that $\sum_{i=i_1+1}^{i_2}|x_{n_2}^i|_i^p < \varepsilon$. (Recall that $\{|x_n^i|_i\}_n$ converges to a^i). Therefore, denoting by $o(1)$ a quantity such that $o(1) \to 0$ as $\varepsilon \to 0$ we have

$$l^p > \|x_{n_1} - x_{n_2}\|^p - \varepsilon = \sum_{i\leq i_1}|x_{n_1}^i - x_{n_2}^i|_i^p$$

$$+ \sum_{i_1<i\leq i_2}|x_{n_1}^i - x_{n_2}^i|_i^p + \sum_{i>i_2}|x_{n_1}^i - x_{n_2}^i|_i^p$$

$$> \sum_{i\leq i_1}l_i^p + \sum_{i_1<i\leq i_2}|x_{n_1}^i|_i^p + \sum_{i>i_2}|x_{n_2}^i|_i^p - o(1)$$

$$\geq \sum_{i\leq i_1}\omega^p(a^i)^p + \sum_{i_1<i\leq i_2}|x_{n_1}^i|_i^p + \sum_{i>i_2}|x_{n_2}^i|_i^p - o(1)$$

$$> \frac{\omega^p}{2}\sum_{i\leq i_1}(|x_{n_1}^i|_i^p + |x_{n_2}^i|_i^p) + \sum_{i_1<i\leq i_2}|x_{n_1}^i|_i^p + \sum_{i>i_2}|x_{n_2}^i|_i^p - o(1)$$

$$> \frac{\omega^p}{2}(\|x_{n_1}\|^p + \|x_{n_2}\|^p) - o(1) = \omega^p - o(1).$$

Since ε is arbitrary we obtain the desired result. $\qquad\square$

Remark 5.3.

(a) An extension of this result to Orlicz sequence spaces and more general direct sum spaces can be found in [Do5].

(b) When X is a finite direct sum of n reflexive Banach spaces $X_1,\ldots,X_n$ a similar (and easier) argument lets us prove that $WCS(X) = \inf\{WCS(X_i) : i = 1,\ldots,n\}$.

We already know that a Banach space X is reflexive if $N(X) > 1$. However we can find nonreflexive Banach spaces, without the Schur property, such that $WCS(X) > 1$.

Example 4: Let X be the space $\ell^1 \otimes \ell^2$ with the norm $\|(x,y)\| = \sqrt{\|x\|_1^2 + \|y\|_2^2}$ and let $\{(x_n,y_n)\}$ be a weakly null sequence which is not norm convergent. Since X satisfies the Opial condition we have that $r_a(\{(x_n,y_n)\}) = \limsup_{n\to\infty}\|(x_n,y_n)\|$. Since $\{x_n\}$ is weakly null and ℓ^1 has the Schur property we know $\{x_n\}$ converges to zero. Therefore $\mathrm{diam}_a(\{(x_n,y_n)\}) = \mathrm{diam}_a(\{y_n\})$ and $\limsup_{n\to\infty}\|(x_n,y_n)\| = \limsup_{n\to\infty}\|y_n\|_2$. Since $WCS(\ell^2) = \sqrt{2}$ we obtain $WCS(X) = \sqrt{2}$ although X is not reflexive.

The following example shows that $WCS(X)$ (and so $N(X)$) can be 1 in a Banach space with normal structure.

Example 5: Let X be the ℓ^2-direct sum of the sequence spaces $\{\ell^n\}_{n\geq 2}$. This space is reflexive and it has normal structure because it is a UCED space (see Remark 2.3 (a)). However, $WCS(X) = \inf\{WCS(\ell^n) : n \geq 2\} = \inf\{2^{1/n} : n \geq 2\} = 1$.

Example 6: Let X be the ℓ^2-direct sum of the sequence spaces $\{\ell^{r_n}\}$ where $r_n = (1+n)/n$. Then $N(X) \leq N(\ell^{r_n})$ for every n which implies $N(X) = 1$ (see Theorem 6.3). However, $WCS(X) = \sqrt{2}$. Thus the condition $WCS(X) > 1$ for a reflexive space does not imply uniform normal structure, although this space has normal structure (Theorem 3.3).

6. Computation of the normal structure coefficients in L^p−spaces

We have seen that it is not difficult to compute $WCS(\ell^p)$. However, for ten years the value of $N(\ell^p)$ and $N(L^p(\Omega))$ for $p \neq 2$ was an open problem. In 1990 the value of these coefficients [Pr1, Do4] was obtained using some convexity inequalities derived from interpolation theory. To simplify the problem we shall start with a lemma by D. Amir [Am].

LEMMA 6.1. *Let X be a reflexive Banach space. Then*

$$N(X) = \inf \left\{ \frac{diam(K)}{r(K)} : K \subset X \;\; finite \; not \; singleton \right\}.$$

Proof. Let A be a convex closed and bounded subset of X. We can assume that $diam(A) = 1$. We choose a positive number $r < r(A)$. Then

$$\bigcap_{x \in A} \overline{B}(x, r) \cap A = \emptyset.$$

Since A and all closed balls are weakly compact there is a finite subset K of A such that

$$\bigcap_{x \in K} \overline{B}(x, r) \cap \; co(K) \subset \bigcap_{x \in K} \overline{B}(x, r) \cap A = \emptyset.$$

Thus $r < r(K)$ which implies $diam(K)/r(K) < 1/r$. Since $r < r(A)$ is arbitrary, we have found a finite subset K of A such that $diam(K)/r(K) \leq diam(A)/r(A)$ and the result follows. $\qquad\square$

A second simplification can be made. We shall prove in the following lemma that we can restrict ourselves to considering only finite sets whose points are all equidistant from a Chebyshev centre.

LEMMA 6.2. *Let X be a Banach space and A a finite subset of X. Then there exists a subset B of A such that*

(i) $r(B) \geq r(A)$, *so* $diam(B)/r(B) \leq diam(A)/r(A)$.
(ii) $\|v - x\| = r(B)$ *for every* $x \in B$ *where* v *is a Chebyshev centre of* B.

Proof. Since A is a finite set, there exists B which is minimal in the family of those nonempty subsets of A which satisfy (i). We shall prove that B satisfies (ii). Since B is finite, $\operatorname{co}(B)$ lies in a finite dimensional space. Thus $Z(B) \neq \emptyset$ and the same applies to any subset of B. Let y_0 be a Chebyshev centre of B and define the set $B_1 = \{x \in B : \|x - y_0\| = r(B)\}$. We shall prove that $r(B_1) \geq r(B)$. Choose a positive real number ε such that $\|x - y_0\| + \varepsilon < r(B)$ for every $x \in B \setminus B_1$. Let y_1 be a Chebyshev centre of B_1 and λ be a real number, $0 < \lambda < 1$, such that $\lambda\|y_0 - y_1\| < \varepsilon/2$. If $r(B_1) < r(B)$ and $x \in B_1$ we have

$$\|x - y_0 + \lambda(y_0 - y_1)\| \leq \lambda\|x - y_1\| + (1-\lambda)\|x - y_0\| \leq \lambda r(B_1) + (1-\lambda)r(B) < r(B).$$

If x belongs to $B \setminus B_1$ one has

$$\|x - y_0 + \lambda(y_0 - y_1)\| \leq \|x - y_0\| + \lambda\|y_0 - y_1\| < r(B) - \varepsilon/2 < r(B).$$

Thus there exists $c < r(B)$ such that $\|x - y_0 + \lambda(y_0 - y_1)\| < c$ for every $x \in B$, contradicting that y_0 is a Chebyshev center of B, because $y_0 + \lambda(y_1 - y_0)$ belongs to $\operatorname{co}(B)$. Since B is minimal satisfying (i) and B_1 also satisfies (i) we have $B = B_1$. Recalling the definition of B_1 it is clear that B satisfies (ii).   □

Now we can compute the values of $N(L^p(\Omega))$.

THEOREM 6.3. *Let (Ω, Σ, μ) be a σ-finite measure space, $1 \leq p < +\infty$ and assume that $L^p(\Omega)$ is infinite dimensional. Then $N(L^p(\Omega)) = \min\{2^{1-1/p}, 2^{1/p}\}$.*

Furthermore $WCS(L^p(\Omega)) = N(L^p(\Omega))$ if either $p \geq 2$ or μ is not purely atomic.

Proof. Let A be a finite set in $L^p(\Omega)$. A subset B of A as in Lemma 6.2 is chosen. If v is the Chebyshev centre of B, we choose real positive numbers $t_1, \ldots, t_n$, with $\sum_{k=1}^n t_k = 1$ and points $x_1, \ldots, x_n$ in B such that $\sum_{k=1}^n t_k x_k = v$. By Lemma II.3. 8 we have

$$(\operatorname{diam}(B))^\beta \geq \sum_{j,k=1}^n t_j t_k \|x_j - x_k\|^\beta \geq 2 \sum_{j=1}^n t_j \left\| x_j - \sum_{k=1}^n t_k x_k \right\|^\beta$$

$$= 2 \sum_{j=1}^n t_j \|x_j - b\|^\beta = 2(r(B))^\beta.$$

Thus $\operatorname{diam}(B) \geq 2^{1/\beta} r(B)$ where $\beta = 1 - 1/p$ if $1 < p \leq 2$ or $\beta = p$ if $2 \leq p < +\infty$. Thus $N(L^p(\Omega)) \geq \min\{2^{1-1/p}, 2^{1/p}\}$.

If $p \geq 2$, by considering the canonical basis of ℓ^p and recalling that ℓ^p is isometrically embedded in every infinite dimensional $L^p(\Omega)$ (see proof of Lemma II.3.9), we have $WCS(L^p(\Omega)) \leq 2^{1/p}$ and the case $p \geq 2$ is concluded. For $p < 2$ consider the set C formed by $r_1, \ldots, r_m$, the first Rademacher functions in $L^p[0,1]$.

It is clear that $\|r_k\| = 1$ and $\|r_k - r_j\| = 2^{1-1/p}$ for every k, j, $k \neq j$. Let r be a mapping in $\mathrm{co}(C)$, that is, $r = \sum_{k=1}^{m} a_k r_k$ where $0 \leq a_k \leq 1$ and $\sum_{k=1}^{n} a_k = 1$. Therefore for $k = 1, \ldots, m$

$$\|r - r_k\| = \left(\int_0^1 |r - r_k|^p d\mu \right)^{1/p} \geq \int_0^1 |r - r_k| d\mu \geq \left| \int_0^1 (r - r_k) r_k d\mu \right| = (1 - a_k).$$

Thus

$$r(C) \geq \max_{k=1,\ldots,m} \{1 - a_k\} = 1 - \min_{k=1,\ldots,m} \{a_k\} \geq 1 - \frac{1}{m}$$

and $\mathrm{diam}(C)/r(C) \leq 2^{1-1/p}(1 - 1/m)^{-1}$.

By discretization of the measure, it is easy to embed isometrically $\mathrm{span}\{r_1, \ldots, r_m\}$ in ℓ^p (actually in $(\mathbb{R}^{2^m}, \|\cdot\|_p)$). Since ℓ^p can be isometrically embedded in $L^p(\Omega)$, we obtain $N(L^p(\Omega)) \leq 2^{1-1/p}(1 - 1/m)^{-1}$ for every $m \in \mathbb{N}$. Since m is arbitrary $N(L^p(\Omega)) = 2^{1-1/p}$ if $p \leq 2$.

If μ is not purely atomic, a Rademacher sequence can be constructed (see Theorem II.3.12). Therefore, it is easy to check that $WCS(L^p(\Omega)) = 2^{1-1/p}$ if $p \leq 2$. $\square$

Remark 6.4.

(a) If μ is purely atomic, $L^p(\Omega)$ is isometrical to ℓ_p. Thus, if $p < 2$ the coefficient $N(L^p(\Omega))$ is $2^{1-1/p}$, strictly les than $WCS(L^p(\Omega)) = 2^{1/p}$. This is an easy example of a reflexive space where the two coefficients are different.

(b) We recall that a Banach space X is said to be *finitely representable* in another Banach space Y if for every finite-dimensional subspace E of X and every $\varepsilon > 0$ there exists a subspace F of Y such that $d(E, F) < 1 + \varepsilon$. It is clear from Lemma 6.2 that the normal structure coefficient of a reflexive Banach space is determined by the finite subsets of the space. Thus, if X and Y are reflexive Banach spaces and X is finitely representable in Y we have $N(X) \geq N(Y)$. On the other hand, it can be proved that $L^p(\Omega)$ is finitely representable in ℓ^p. Indeed, if E is an n-dimensional subspace of $L^p(\Omega)$ and $\{f_1, f_2, \ldots, f_n\}$ is a normalized basis of E, with basic constant c, for every $\varepsilon > 0$ we can find simple functions $\{s_1, s_2, \ldots, s_n\}$ such that $\|f_k - s_k\| < \varepsilon/nc(2 + \varepsilon)$ for $k = 1, \ldots, n$. If $f = \sum_{k=1}^n a_k f_k$, we define $Tf = \sum_{k=1}^n a_k s_k$. Then T maps isomorphically E onto $\mathrm{span}\{s_1, \ldots, s_n\}$ and $\|T\|\|T^{-1}\| < 1 + \varepsilon$. Since $\mathrm{span}\{s_1, \ldots, s_n\}$ can be embedded isometrically in ℓ^p (by discretization of the measure), there exists a subspace F of ℓ^p such that $d(E, F) < 1 + \varepsilon$. Since ℓ^p is isometrically embedded in $L^p(\Omega)$, the finite representability of $L^p(\Omega)$ in ℓ^p shows that the coefficients $N(L^p(\Omega))$ and $N(\ell^p)$ must therefore be equal, as checked in Theorem 6.3. Since $WCS(L^p([0,1]))$ and $WCS(\ell^p)$ are different for $p < 2$, it is clear that the weakly convergent sequence coefficient is not determined by the finite subsets of the space.

The above considerations are also useful to obtain an upper bound for $N(X)$.

COROLLARY 6.5. *Let X be an infinite dimensional Banach space. Then*

$$N(X) \leq \sqrt{2}.$$

Proof. From Dvoretzky's Theorem (see [Pi, Theorem 14.1, page 41] or [FLM]) we know that ℓ^2 is finitely representable in every infinite dimensional Banach space X. From Theorem 6.3 and Remark 6.4 (b) we have $N(X) \leq N(\ell_2) = \sqrt{2}$.   □

Chapter VII

Fixed Point Theorems in the Absence of Normal Structure

In Chapter VI we studied the f.p.p. as a property which is implied by normal structure. However, there are Banach spaces without normal structure which have the f.p.p. For instance, $\ell^{p,\infty}$ does not have normal structure (Example VI.2) but this space has the f.p.p. This fact can be proved, for instance, checking that the Banach-Mazur distance between $\ell^{p,\infty}$ and ℓ^p is $2^{1/p}$ and applying a stability result in [By3].

Stanisław Mazur (1905–1981) was born in Lwów (Poland) on 1 January. He studied mathematics in Lwów and Paris. He took his doctorate at Jan Kazimierz University in Lwów and began work as an assistant of Hugo Steinhaus. During the period 1939–1941 of Soviet rule, he held the chair of Geometry. During the German occupation he worked as a shop assistant. After the German withdrawal from Lwów, he returned to his former position, but for two years devoted himself to organizing the repatriation of the Polish population from the Soviet Union. After returning to Poland in 1946 he became a Professor of the newly established University of Łódź. After two years he transferred to the University of Warsaw, where he remained until his retirement

Stanisław Mazur had wide mathematical interests. His first publications concerned the theory of summability, and he returned to this field in later periods. Under the influence of Stefan Banach, he took up functional analysis. Many of the joint results of Banach and Mazur appeared in Banach's Monograph "Théorie des opérations linéaires". One of them is the theorem of universality of the space $C([0,1])$ for separable Banach spaces.

Mazur is universally recognized as the inventor of geometrical methods in functional analysis. He is the author of the theorem on weak closedness of closed convex sets, and of the result on the character of the set of points of differentiability of a convex functional. He, independently of J. von Neumann, introduced the concept of locally convex spaces, and, jointly with Orlicz, systematically studied completely metrizable locally convex spaces.

Stanisław Mazur was the teacher of many Polish mathematicians. With Orlicz, he reconstructed the Polish school of functional analysis after the Second World War. His charismatic personality and his talent as a teacher attracted many young mathematicians, and his seminars in the Institute of Mathematics of the Polish Academy of Sciences and at the University of Warsaw were for many years the inspiration of research in functional analysis in Warsaw.

NUS spaces can also fail to have normal structure. The space $\ell^{p,\infty}$ is again an example of this assertion. Since $\ell^{p,1}$ is 2-UC we know that $\ell^{p,\infty}$ is 2-US and so NUS. However, it can be proved that NUS spaces have the f.p.p. [Ga]. These facts suggest the definition of a new coefficient which assures the f.p.p. in a more general setting than the normal structure coefficient does. In Section 1 we prove a "classic" result in metric fixed point theory: Goebel-Karlovitz's lemma [Go2, Ka2]. This lemma is used to prove another important result in the theory: Lin's lemma [Ln1]. In Section 2 we define a new coefficient for Banach spaces, denoted by $M(X)$, and we prove that X has the weak fixed point property if $M(X) > 1$. This coefficient satisfies $M(X) \geq WCS(X)$, the inequality being strict in certain spaces, for instance ℓ^p or $\ell^{p,q}$, $1 < p < \infty$, $1 \leq q \leq \infty$. We also obtain a lower bound for $M(X)$ where either the k-US modulus or the NUS modulus plays the role of the US-modulus in Theorem VI.4.1. The Opial modulus of the dual space also gives us a lower bound for $M(X)$.

1. Goebel-Karlovitz's lemma and Lin's lemma

We are going to prove two basic results in metric fixed point theory: Goebel-Karlovitz's lemma and Lin's lemma. We need a previous definition.

DEFINITION 1.1. *Let (X, d) be a metric space and T a mapping from X into X. A sequence $\{x_n\}$ is called an approximated fixed point sequence for T if $d(x_n, Tx_n) \to 0$ as $n \to \infty$.*

If we assume that C is a convex bounded closed subset of a Banach space X and $T : C \to C$ a nonexpansive mapping, it is easy to prove that an approximated fixed point sequence exists in C.

PROPOSITION 1.2. *Let K be a weakly compact convex subset of a Banach space X, and $T : K \to K$ be a nonexpansive mapping. Assume that K is minimal for T, that is, no closed convex bounded proper subset of K is invariant for T. If $\{x_n\}$ is an approximated fixed point sequence in K, then*

$$Z_a(\{x_n\}, K) = K.$$

Proof. Let

$$Z_{a,\varepsilon}(\{x_n\}, K) = \{y \in K : \limsup_{n\to\infty} \|x_n - y\| \le r_a(\{x_n\}, K) + \varepsilon\}.$$

It is easy to check that $Z_{a,\varepsilon}(\{x_n\}, K)$ is nonempty, closed, convex and invariant for T. Thus $Z_{a,\varepsilon}(\{x_n\}, K) = K$ and $Z_a(\{x_n\}, K) = \bigcap_{\varepsilon>0} Z_{a,\varepsilon}(\{x_n\}, K) = K$. $\square$

LEMMA 1.3 (GOEBEL-KARLOVITZ). *Let K be a weakly compact convex subset of a Banach space X, and $T : K \to K$ be a nonexpansive mapping. Assume that K is minimal for T and $\{x_n\}$ is an approximated fixed point sequence for T. Then*

$$\lim_{n\to\infty} \|y - x_n\| = \operatorname{diam}(K)$$

for every $y \in K$.

Proof. From the proof of Theorem VI.1.5 we know that $K = Z(K)$ and so K is a diametral set. We claim that $\limsup_{n\to\infty} \|y - x_n\| = \operatorname{diam}(K)$ for every $y \in K$. Indeed, we assume that a vector $y \in K$ exists such that $\limsup_{n\to\infty} \|y - x_n\| < \operatorname{diam}(K)$. We let $r = \limsup_{n\to\infty} \|y - x_n\|$ and $d = \operatorname{diam}(K)$ and consider the family $\{\overline{B}(z, (r + d)/2) \cap K : z \in K\}$. Choose any positive $\varepsilon < (d - r)/2$. From Proposition 1.2 we know that $\limsup_{n\to\infty} \|x_n - z\| = r$ for any $z \in K$. Thus, for any finite subset $\{z_1, \ldots, z_k\}$ of K an integer N exists such that $\|x_N - z_i\| \le r + \varepsilon = (r + d)/2$ for $i = 1, \ldots, k$. Thus x_N belongs to $\bigcap_{i=1}^{k} \overline{B}(z_i, (r + d)/2)$. The weak compactness of K implies the existence of $x_0 \in \bigcap_{z\in K} \overline{B}(z, (r + d)/2) \cap K$ and this point is not diametral because

$$\sup_{z\in K} \|z - x_0\| < \frac{r + d}{2} < d = \operatorname{diam}(K).$$

This contradiction proves the claim. If $\liminf_{n\to\infty} \|y - x_n\| < \operatorname{diam}(K)$ for some $y \in K$ there is a subsequence $\{y_n\}$ of $\{x_n\}$ such that $\limsup_{n\to\infty} \|y_n - y\| = \liminf_{n\to\infty} \|x_n - y\| < \operatorname{diam}(K)$, contradicting the claim, because every subsequence of an approximated fixed point sequence is also an approximated fixed point sequence. $\square$

Let X be a Banach space. We denote $\ell^\infty(X)$ (respectively $c_0(X)$) the linear space of all bounded sequences (respectively all sequences convergent to zero) in X. By $[X]$ will denote the quotient space $\ell_\infty(X)/c_0(X)$ endowed with the norm $\|[z^n]\| = \limsup_{n\to\infty} \|z^n\|$ where $[z^n]$ is the equivalent class of $(z^n) \in \ell_\infty(X)$. By identifying $x \in X$ with the class $[(x, x, \ldots)]$ we can consider X as a subset of $[X]$. If K is a subset of X we can define the set $[K] = \{[z^n] \in [X] : z^n \in K$ for every $n \in \mathbb{N}\}$. If T is a mapping from K into K, then $[T] : [K] \to [K]$ given by $[T]([z^n]) = [Tx^n]$ is a well defined mapping.

LEMMA 1.4 (LIN). *Let X be a Banach space and K be a weakly compact convex subset of X. Let $T : K \to K$ a nonexpansive map and suppose K is a minimal invariant set for K. If $[W]$ is a nonempty closed convex subset of $[K]$ which is invariant under $[T]$ then*

$$\sup \{ \| [w^n] - [x] \| : [w^n] \in [W] \} = \operatorname{diam}(K)$$

for every $x \in K$.

Proof. We claim that $\limsup_{m \to \infty} \| [w^n]_m - [x] \| = \operatorname{diam}(K)$ for every $x \in K$, $\{ [w^n]_m \}$ being an approximated fixed point sequence for $[T]$ in $[W]$, and this claim clearly proves the lemma, because $[T]$ is also nonexpansive and we can find a sequence with this property in $[W]$. We denote a representative of the n-th element of the sequence $\{ [w^n]_m \}$ as w_m^n and we write $d = \limsup_{m \to \infty} \| [w^n]_m - [x] \|$ and $\delta_m = \| [w^n]_m - [T][w^n]_m \|$. Thus $\lim_{m \to \infty} \delta_m = 0$. We shall prove $d = \operatorname{diam}(K)$. Since we obviously have $d \leq \operatorname{diam}(K)$ we only need to prove the inequality $d \geq \operatorname{diam}(K)$. To this end we construct a point $[w^k] \in [K]$ such that

(a) $[T][w^k] = [w^k]$.
(b) $\| [w^k] - [x] \| \leq d$.

Thus Lemma 1.3 will imply

$$\operatorname{diam}(K) = \lim_{k \to \infty} \| w^k - x \| = \| [w^k] - x \| \leq d.$$

Choose a sequence $\{ \varepsilon_k \} \to 0$. For a fixed $k \in \mathbb{N}$ a positive integer m_k exists such that $\| [w^n]_m - x \| \leq d + \varepsilon_k$ if $m \geq m_k$. Since

$$\limsup_{n \to \infty} \| w_{m_k}^n - x \| \leq d + \varepsilon_k \ \text{ and } \ \limsup_{n \to \infty} \| w_{m_k}^n - T w_{m_k}^n \| = \delta_{m_k}$$

we can choose a large enough n_k, such that

$$\| w_{m_k}^{n_k} - x \| \leq d + 2\varepsilon_k \ \text{ and } \ \| w_{m_k}^{n_k} - T w_{m_k}^{n_k} \| \leq \delta_{m_k} + \varepsilon_k.$$

Now consider the sequence $[w^k] = [w_{m_k}^{n_k}] \in [K]$. It is clear that $[w^k]$ satisfies (a) and (b). $\qquad\qquad\square$

2. The coefficient $M(X)$ and the fixed point property

In this section we are going to introduce a new coefficient in Banach spaces which yields a new fixed point theorem. As we shall see, this theorem enables us to prove the existence of a fixed point in Banach spaces without normal structure. First we need to define a uniparameter family of coefficients.

DEFINITION 2.1. *Let X be a Banach space. For any nonnegative number a we define the coefficient*

$$R(a, X) = \sup\{\liminf_{n\to\infty} \|x_n + x\|\}$$

where the supremum is taken over all $x \in X$ with $\|x\| \le a$ and over all weakly null sequences in $\overline{B}(0, 1)$ such that $\lim_{n,m\,;n\neq m} \|x_n - x_m\| \le 1$.

THEOREM 2.2. *Let X be a Banach space and assume that for some $a \ge 0$ we have $R(a, X) < 1 + a$. Then X has the weak fixed point property.*

Proof. We follow an argument similar to that in [Ga]. If we assume that X fails to have the w.f.p.p., we can find a weakly compact and convex subset K of X such that $\operatorname{diam}(K) = 1$ and K is minimal invariant for a nonexpansive mapping T and a weakly null approximated fixed point sequence $\{x_n\}$ in K. We shall consider this sequence as the point (x^n) in $\ell^\infty(X)$ and define the set

$$[W] = \{[z^n] \in [K] : \|[z^n] - [x^n]\| \le 1 - t \text{ and } \limsup_{n\to\infty}\limsup_{m\to\infty} \|z^n - z^m\| \le t\}$$

where $t = 1/(1 + a)$. It is easy to check that $[W]$ is a closed, convex and $[T]$-invariant set. Furthermore, $[W]$ is nonempty because it contains $[tx^n]$. Therefore, from Lemma 1.4 we know that

$$\sup\{\|[w^n] - [x]\| : [w^n] \in [W]\} = 1$$

for every $x \in K$. We take $[z^n] \in [W]$ and choose a sequence $\{z_n\}$ in K such that $\{z_n\} \in [z^n]$. Let $\{y_n\}$ be a weakly convergent subsequence of $\{z_n\}$ such that $\limsup_{n\to\infty} \|z_n\| = \lim_{n\to\infty} \|y_n\|$ and $\lim_{n,m\,;n\neq m} \|y_n - y_m\|$ exists. In this way we have

$$\lim_{n,m\,;n\neq m} \|y_n - y_m\| = \limsup_{n\to\infty}\limsup_{m\to\infty} \|y_n - y_m\| \le \limsup_{n\to\infty}\limsup_{m\to\infty} \|z_n - z_m\| \le t.$$

We denote the weak limit of $\{y_n\}$ by y. For every $n \in \mathbb{N}$ we have $\|y_n - y\| \le \liminf_{m\to\infty} \|y_n - y_m\|$. Hence

$$\limsup_{n\to\infty} \|y_n - y\| \le \limsup_{n\to\infty}\limsup_{m\to\infty} \|y_n - y_m\| \le t.$$

A positive η can be chosen such that $\eta R(a, X) < 1 - R(a, X)/(1 + a)$. For a large enough n, we have $\|y_n - y\| \le t + \eta$. Furthermore $\|y\| \le \liminf_{n\to\infty} \|y_n - x_n\| \le 1 - t$. Hence

$$\left\| \frac{y_n}{t + \eta} \right\| = \left\| \frac{y_n - y}{t + \eta} + \frac{y}{t + \eta} \right\| \le R\left(\frac{1 - t}{t}, X \right) = R(a, X).$$

Thus $\limsup_{n\to\infty} \|z_n\| = \lim_{n\to\infty} \|y_n\| \le R(a, X)(t + \eta) < 1$ which is a contradiction because $0 \in K$. $\square$

The following stability result, similar to those in Theorems VI.2.8 and VI.3.5, can be proved by a straightforward argument.

THEOREM 2.3. *Let X and Y be isomorphic Banach spaces. Then $R(a,Y) \leq d(X,Y)R(a,X)$ for every nonnegative number a.*

DEFINITION 2.4. *Let X be a Banach space. We define the coefficient $M(X)$ as*

$$\sup\left\{\frac{1+a}{R(a,X)} : a \geq 0\right\}.$$

The following result is a direct consequence of Theorems 2.2 and 2.3.

THEOREM 2.5. *Let X be a Banach space. If $M(X) > 1$ then X has the w.f.p.p. If Y is another Banach space which is isomorphic to X and if $d(X,Y) < M(X)$, then Y has the w.f.p.p.*

Remark 2.6.

(a) It must be noted that $WCS(X) = 1/R(0,X)$ (see Lemma VI.3.8). Thus $M(X) \geq WCS(X)$. We can show examples where this inequality is strict:

(i) Assume $X = \ell^2$. To compute $R(a, \ell^2)$, we consider a weakly null sequence $\{x_n\}$ in the unit ball of ℓ^2 and a vector x with $\|x\| \leq a$. Taking a subsequence and using standard arguments we can construct sequences $\{u_n\}$ and $\{v_n\}$ such that $\lim_{n\to\infty}\|x - u_n\| = 0$, $\lim_{n\to\infty}\|x_n - v_n\| = 0$ and $\operatorname{supp}(u_n) \cap \operatorname{supp}(v_n) = \emptyset$ for every $n \in \mathbb{N}$. Furthermore, the condition $\lim_{n,m\,;n\neq m}\|x_n - x_m\| \leq 1$ implies $\limsup\|v_n\| \leq 1/\sqrt{2}$ because $WCS(\ell^2) = \sqrt{2}$ (see Lemma VI.3.8 (d)). Then we have

$$\liminf_{n\to\infty}\|x + x_n\|^2 = \liminf_{n\to\infty}\|u_n + v_n\|^2 = \liminf_{n\to\infty}(\|u_n\|^2 + \|v_n\|^2) = a^2 + 1/2.$$

Now, it is easy to check that $R(a, \ell^2) = \sqrt{a^2 + 1/2}$ and the function $(1 + a)/\sqrt{a^2 + 1/2}$ attains its maximum at $a = 1/2$ with value $M(\ell^2) = \sqrt{3}$. In the same way we can check that

$$M(\ell^p) = \left(1 + 2^{\frac{1}{p-1}}\right)^{\frac{p-1}{p}}$$

for every $p \in (1, +\infty)$. Note that for all $p \in (1, +\infty)$ we have $M(\ell^p) > WCS(\ell^p)$.

(ii) Assume $X = c_0$. An argument similar as that in (i) proves that $R(a, c_0) = 1 + a$ if $a < 1$ and $R(a, c_0) = a$ if $a \geq 1$. Thus $M(c_0) = 2$ and c_0 has the w.f.p.p. although c_0 does not have weak normal structure.

(iii) On the other hand, for Bynum's space $\ell^{p,\infty}$ we have $M(\ell^{p,\infty}) = 2^{1-1/p}$. Indeed, let $\{x_n\}$ be a weakly null sequence in $\overline{B}(0,1)$ and x a vector in $\overline{B}(0,a)$. As in (i), we can construct sequences $\{u_n\}$ and $\{v_n\}$ such that $\lim_{n\to\infty}\|x - u_n\| = 0$, $\lim_{n\to\infty}\|x_n - v_n\| = 0$ and $\operatorname{supp}(u_n) \cap \operatorname{supp}(v_n) = \emptyset$ for every $n \in \mathbb{N}$. It is clear that $\lim_{n\to\infty}\|(u_n + v_n)^+ - (x + x_n)^+\| = \lim_{n\to\infty}\|(u_n + v_n)^- - (x + x_n)^-\| = \lim_{n\to\infty}\|u_n^+ - x^+\| = \lim_{n\to\infty}\|u_n^- - x^-\| = \lim_{n\to\infty}\|v_n^+ - x_n^+\| = \lim_{n\to\infty}\|v_n^- - x_n^-\| = 0$. Since $\operatorname{supp}(u_n) \cap \operatorname{supp}(v_n) = \emptyset$, we have

$$\lim_{n\to\infty}\|(x + x_n)^+\| = \lim_{n\to\infty}\|(u_n + v_n)^+\| = \lim_{n\to\infty}\|u_n^+ + v_n^+\| \leq (1 + a^p)^{1/p}$$

and similarly

$$\lim_{n\to\infty} \|(x+x_n)^-\| = \lim_{n\to\infty} \|(u_n+v_n)^-\| = \lim_{n\to\infty} \|u_n^- + v_n^-\| \le (1+a^p)^{1/p}.$$

Thus $R(a, \ell^{p,\infty}) \le (1+a^p)^{1/p}$ and it is easy to check that this bound is attained when the sequence $\{e_{n+1}\}$ and the vector ae_1 are considered. The maximum of $(1+a)/(1+a^p)^{1/p}$ is now attained at $a = 1$ and the corresponding value is $M(\ell^{p,\infty}) = 2^{1-1/p}$.

(b) The computation of $M(X)$ for ℓ^2, ℓ^p or $\ell^{p,\infty}$ gives us stability results better than those which can be obtained using $WCS(X)$. Furthermore, it must be noted that these new stability results are also better (see [Do8]) than the results obtained for ℓ^2 in [JL] and for ℓ^p in [Kh2, Pr1, BoS].

We shall prove that $M(X)$ can be bounded from below by means of the modulus of k-US and by means of the modulus of NUS in a similar way to $WCS(X)$ being bounded using the modulus of US (Theorem VI.4.1).

THEOREM 2.7. *Let X be a Banach space and let*

$$\beta = \inf\left\{1 + \beta_X^k(s) - \frac{s}{2k} : s \in [0,1]\right\}.$$

Then $M(X) > (1+2k)/(1+2k\beta)$. In particular X has the fixed point property if $\lim_{t\to 0+} \beta(t)/t < 1/2k$.

Proof. We first consider the case when $a \le 2k$. Let η be an arbitrary positive number, $\{x_n\}$ a weakly null sequence in $\overline{B}(0,1)$ and $x \in X$ such that $\|x\| = r \le a$, $\lim_{n\to\infty} \|x_n + x\|$ exists and $R(a, X) < \lim_{n\to\infty} \|x + x_n\| + \eta/2$. We can assume that $R(a, X) < \|x + x_n\| + \eta$ for every n. A number $t \in [0, r/2k]$ can be chosen such that

$$1 - \frac{t}{r} + \beta_X^k\left(\frac{2kt}{r}\right) < \beta + \eta.$$

If $\{x_n\} \to 0$ it is clear that

$$R(a, X) - \frac{\eta}{2} \le \liminf_{n\to\infty} \|x + x_n\| \le \liminf_{n\to\infty} r \left\|\frac{x}{r} + \frac{tx_n}{r}\right\| + (1-t)$$

$$\le r\left(1 + \beta_X^k\left(\frac{2kt}{r}\right)\right) + (1-t) \le 1 + r\beta + r\eta \le 1 + a\beta + a\eta.$$

If $\{x_n\}$ is not norm convergent, we can assume that $\{x_n\}$ is a basic sequence ([LT1, page 5]). We choose norm one functionals $x_n^* \in X^*$ such that $x_n^*(x+tx_n) = \|x + tx_n\|$. Taking subsequences we can also assume $|x_n^*(x_m)| \le \eta$ if $n \ne m$. Writing $y_n = (x_{2n} - x_{2n+1})/2$ we know that the vectors $y_1, y_2, \ldots, y_k$ are linearly independent. Therefore a normalized vector $y = \sum_{n=1}^{k} \alpha_n y_n$ exists such that

$$\frac{1}{2}\left(\left\|\frac{x + 2tky}{r}\right\| + \left\|\frac{x - 2tky}{r}\right\|\right) \le 1 + \beta_X^k\left(\frac{2tk}{r}\right) + \eta.$$

Assume $|\alpha_m| = \max\{|\alpha_n| : n = 1, \ldots, k\}$. It is clear that $\alpha_m \geq 1/k$. Therefore we have

$$
\begin{aligned}
R(a,X) - \eta &\leq \frac{1}{2}(\|x + x_{2m}\| + \|x + x_{2m+1}\|) \\
&\leq \frac{1}{2}(\|x + tx_{2m}\| + \|x + tx_{2m+1}\|) + (1 - t) \\
&= \frac{1}{2}r\left(\left\|\frac{x}{r} + \frac{tx_{2m}}{r}\right\| + \left\|\frac{x}{r} + \frac{tx_{2m+1}}{r}\right\|\right) + (1 - t) \\
&= \frac{1}{2}r\left[x_{2m}^*\left(\frac{x}{r} + \frac{tx_{2m}}{r}\right) + x_{2m+1}^*\left(\frac{x}{r} + \frac{tx_{2m+1}}{r}\right)\right] + (1 - t) \\
&\leq \frac{1}{2}r\left[x_{2m}^*\left(\frac{x}{r}\right) + x_{2m}^*\left(\frac{2ty}{r\alpha_m}\right) + x_{2m+1}^*\left(\frac{x}{r}\right) - x_{2m+1}^*\left(\frac{2ty}{r\alpha_m}\right)\right] \\
&\quad + 2\eta \sum_{n=1}^{k} \frac{|\alpha_n|}{|\alpha_m|} + (1 - t) \\
&\leq \frac{1}{2}r\left(\left\|\frac{x}{r} + \frac{2ty}{r\alpha_m}\right\| + \left\|\frac{x}{r} - \frac{2ty}{r\alpha_m}\right\|\right) + (1 - t) + 2\eta k \\
&\leq r\left(1 + \beta_X^k\left(\frac{2tk}{r}y\right) - \frac{t}{r}\right) + \eta(2k + 1) + 1 \leq r\beta + 1 + 2\eta(k + 1 + r).
\end{aligned}
$$

Letting $\eta \to 0$, we obtain $R(a, X) \leq 1 + \beta a$. If $a \geq 2k$ we have

$$
\|x + x_n\| \leq (r - 2k) + \left\|\frac{2kx}{r} + x_n\right\|.
$$

Applying the above argument for the sequence $2kx/r + x_n$ we have $R(a, X) \leq (a - 2k) + 1 + 2k\beta = a - (2k - 1) + 2k\beta$. For $a = 2k$ we obtain $M(X) \geq (1 + 2k)/(1 + 2k\beta)$.
$\square$

COROLLARY 2.8. *Let X be a k-US Banach space. Then X has the f.p.p.*

THEOREM 2.9. *Let X be a reflexive Banach space and let*

$$
\Gamma = \inf\left\{1 + \Gamma_X(s) - \frac{s}{2} : s \in [0, 1]\right\}.
$$

Then $R(a, X) \leq 1 + a\Gamma$ if $a \leq 2$ and $R(a, X) \leq a + 2\Gamma - 1$ if $a \geq 2$. In particular, $M(X) \geq 3/(1 + 2\Gamma)$ and $M(X) > 1$ if $\Gamma_X'(0) < 1/2$.

Proof. The statement is obvious if $a = 0$. Assume $2 \geq a > 0$. Let $\{x_n\}$ be a weakly null sequence in $\overline{B}(0, 1)$ and $x \in X$ be a vector such that $\|x\| = r \leq a$. Taking subsequences we can assume that $\lim_{n \to \infty} \|x_n + x\|$ exists. For an arbitrary positive number η, a number $t \in [0, r/2]$ can be chosen such that

$$
1 - \frac{t}{r} + \Gamma_X\left(\frac{2t}{r}\right) < \Gamma + \eta.
$$

With these assumptions we have

$$\|x + x_n\| = r\left\|\frac{x}{r} + \frac{x_n}{r}\right\| \le r\left\|\frac{x}{r} + \frac{t}{r}x_n\right\| + (1 - t).$$

If $\{x_n\} \to 0$, and using $r\Gamma \le a\Gamma$ (because $\Gamma \ge 1/2 > 0$), it is clear that

$$\liminf_{n\to\infty}\|x + x_n\| \le r\left(1 + \Gamma_X\left(\frac{2t}{r}\right)\right) + (1 - t) \le r\eta + r\Gamma + 1 \le 1 + a\Gamma + a\eta.$$

If $\{x_n\}$ does not converge to zero, we can assume that $\{x, x_n\}$ is a basic sequence with arbitrary basic constant $c > 1$ ([LT1, page 5]). Hence, we have

$$\|x + tx_n\| = \frac{1}{2}\|2x + 2tx_n\| \le \frac{1}{2}(\|x\| + \|x + 2tx_n\|) \le \frac{1}{2}(c\|x - 2tx_n\| + \|x + 2tx_n\|).$$

Again taking subsequences we can assume

$$\frac{1}{2}\left(\left\|\frac{x}{r} + \frac{2t}{r}x_n\right\| + \left\|\frac{x}{r} - \frac{2t}{r}x_n\right\|\right) - 1 \le \Gamma_X\left(\frac{2t}{r}\right) + \eta.$$

Thus

$$\|x + x_n\| \le \frac{rc}{2}\left(\left\|\frac{x}{r} - \frac{2t}{r}x_n\right\| + \left\|\frac{x}{r} + \frac{2t}{r}x_n\right\|\right) + (1 - t)$$

$$\le rc\left[1 + \Gamma_X\left(\frac{2t}{r}\right) + \eta\right] + (1 - t)$$

$$\le c\left[r\left(1 + \Gamma_X\left(\frac{2t}{r}\right) + \eta - \frac{t}{r}\right) + 1\right] + (c - 1)t$$

$$\le c(1 + r\Gamma + 2r\eta) + (c - 1)a \le c(1 + a\Gamma + 2a\eta) + (c - 1)a.$$

Hence

$$R(a, X) \le c(1 + a\Gamma + 2a\eta) + (c - 1)a.$$

Since $c > 1$ and $\eta > 0$ are arbitrary we obtain $R(a, X) \le 1 + a\Gamma$. If $a \ge 2$ we have

$$\|x + x_n\| \le (r - 2) + \left\|\frac{2x}{r} + x_n\right\|.$$

Applying the above result for the sequence $2x/r + x_n$ we have $R(a, X) \le (a - 2) + 1 + 2\Gamma = a - 1 + 2\Gamma$. For $a = 2$ we obtain $M(X) \ge 3/(1 - 2\Gamma)$. Finally, if $\Gamma'_X(0) < 1/2$ it is clear that $\Gamma < 1$. $\square$

COROLLARY 2.10. *Let X be an NUS Banach space. Then X has the f.p.p.*

To conclude this chapter, we shall show that the modulus of Opial of the dual space can also be used to obtain a lower bound for $M(X)$.

THEOREM 2.11. *Let X be a Banach space such that the unit ball of X^* is weakly* sequentially compact. If $c_1 \in (0, 1)$ satisfies $r_{X^*}(c_1) > 0$, then*

$$R(a, X) \le \max\left\{1 + ac_1, a + \frac{1}{1 + r_{X^*}(c_1)}\right\}.$$

In particular $R(a, X) < 1 + a$ and $M(X) > 1$ if $r_{X^*}(1) > 0$.

Proof. We assume that $\{x_n\}$ is a weakly null sequence in $\overline{B}(0,1)$ and that $x \in X$ satisfies $\|x\| \le a$. We choose norm one functionals $z_n^* \in X^*$ such that $z_n^*(x+x_n) = \|x+x_n\|$. Taking subsequences we can assume that $\{z_n^*\}$ is weakly convergent to a point, say z^*, and that $\lim_{n\to\infty} \|z_n^* - z^*\| = d$ exists. Let $\|z^*\| = c$. If $c \le c_1$ we have

$$\liminf_{n\to\infty} \|x + x_n\| = \liminf_{n\to\infty} z_n^*(x + x_n) \le z^*(x) + \liminf_{n\to\infty} z_n^*(x_n) \le ca + 1 \le 1 + ac_1.$$

If $c > c_1$ we claim that $d \le 1/(1 + r_{X^*}(c))$. Indeed, if $d > 1/(1 + r_{X^*}(c))$ we can choose $\alpha > 1$ satisfying $1/d < \alpha < 1 + r_{X^*}(c)$. Since

$$\|\alpha(z_n^* - z^*)\| > \left\| \frac{z_n^* - z^*}{d} \right\|,$$

we have $\lim_{n\to\infty} \|\alpha(z_n^* - z^*)\| \ge 1$. Thus

$$\alpha = \|\alpha z_n^*\| = \|\alpha(z_n^* - z^*) + \alpha z^*\| = \lim_{n\to\infty} \|\alpha(z_n^* - z^*) + \alpha z^*\| \ge 1 + r_{X^*}(\alpha c) \ge 1 + r_{X^*}(c)$$

which is a contradiction. So we have $d \le 1/(1 + r_{X^*}(c))$ and

$$\liminf_{n\to\infty} \|x + x_n\| = \lim_{n\to\infty} z^*(x + x_n) + \liminf_{n\to\infty}(z_n^* - z^*)(x + x_n)$$

$$= z^*(x) + \liminf_{n\to\infty}(z_n^* - z^*)(x_n) \le a + \frac{1}{1 + r_{X^*}(c_1)}.$$

Now, the last assertion is clear from the continuity of r_{X^*} (Theorem V.3.5(e)). $\square$

COROLLARY 2.12. *Let X be a reflexive Banach space. If X^* satisfies the uniform Opial condition, then X has the f.p.p.*

Chapter VIII

Uniformly Lipschitzian Mappings

Assume that M is a metric space and $T : M \to M$ is nonexpansive. Clearly T and all iterate mappings T^n are Lipschitzian with constant $k = 1$.

> **Rudolf Otto Sigmund Lipschitz** (1832–1903) was born in Königsberg (Germany) on 14 May. In 1847 he entered the University of Königsberg, where one of his teachers was Franz Neumann. He continued his studies at the University of Berlin, chiefly under Lejeune Dirichlet. He taught at schools in Königsberg and Elbinc, before becoming a lecturer at the University of Berlin in 1857. He was appointed Professor of Mathematics at the University of Bonn in 1864 and remained there for the rest of his career, so contented with his work and life there that he turned down an invitation to become a professor at the more prestigious University of Göttingen. He died in Bonn on 7 October.
>
> Lipschitz did extensive work in number theory, Fourier series, the theory of Bessel functions, differential equations, the calculus of variations, geometry and mechanics. Among his more specific contributions to mathematical knowledge, several stand out. His work in basic analysis provided a condition now known as the Lipschitz condition, subsequently of great importance in proofs of existence and uniqueness, as well as in approximation theory and constructive function theory.
>
> Partly because he spread himself so wide, Lipschitz's star does not shine so brightly as some others in the mathematical firmament; but he was one of the most industrious and most technically proficient of nineteenth-century mathematicians.

If we consider mappings satisfying the same property for $k > 1$ we obtain a natural generalization of the nonexpansive mappings. An interesting question now appears: How small must k be to assure the existence of a fixed point for T? The first results in this direction were given by Goebel and Kirk [GK2] who stated a relationship between the existence of a fixed point for uniformly Lipschitzian mappings and the Clarkson modulus of convexity. A more general approach is proposed by Lifshitz [L]. He defines a coefficient $\kappa(M)$ in a metric space M, which will be called the Lifshitz characteristic, and proves a fixed point theorem for k-uniformly Lipschitzian mappings when $k < \kappa(M)$. A considerable number of papers have appeared in the last twenty years on this subject (see for instance [Gr1], [Gr2], [Zh1], [T], [DT] or [DX]). In this chapter we shall study this problem. In Section 1 we show that when M is a closed, convex and bounded subset of a Banach space, then the Clarkson modulus gives a bound such that T has a fixed point if k is smaller that this bound. We define the Lifshitz characteristic in a general setting and we show its application to obtain a fixed point theorem. In Section 2 we study the relationships between the Lifshitz characteristic and both the Clarkson modulus and the normal structure coefficient, and we compute its value in Hilbert spaces. In Section 3 we show how the normal structure coefficient can also be used to obtain a fixed point theorem for uniformly Lipschitzian mappings.

1. Lifshitz characteristic and fixed points

DEFINITION 1.1. *Let (X, d) and (Y, d) be metric spaces. A mapping $T : X \to Y$ is called uniformly Lipschitzian if there exists a constant k such that*

$$d(T^n x_1, T^n x_2) \le k d(x_1, x_2)$$

for any points x_1 and x_2 in X and any positive integer n.

Let T be a nonexpansive mapping from a subset C of a Banach space X into X. Assume that Y is another Banach space isomorphic to X and let $f : X \to Y$ be an isomorphism. Then the mapping $f \circ T \circ f^{-1}$ from $f(C)$ into Y is Lipschitzian with constant $\|f\| \|f^{-1}\|$. Furthermore all iterated mappings are also Lipschitzian with the same constant. Indeed:

$$\|f \circ T^n \circ f^{-1}(x) - f \circ T^n \circ f^{-1}(y)\|$$
$$\le \|f\| \|T^n \circ f^{-1}(x) - T^n \circ f^{-1}(y)\| \le \|f\| \|f^{-1}\| \|x - y\|.$$

In the reverse direction, it can be proved that if T is uniformly Lipschitzian, then T is nonexpansive with respect to an equivalent metric [GK1, page 170]. Thus the problem of studying the stability of the fixed point property under isomorphisms or renormings drives us to study the existence of a fixed point for uniformly Lipschitzian mappings. We begin this chapter with a theorem [GK2] which states a relationship between the existence of a fixed point for uniformly Lipschitzian mappings and the Clarkson modulus of convexity. We do not include the proof because in Theorem 2.1 we shall establish a more general result.

THEOREM 1.2. *Let X be a uniformly convex Banach space with modulus of convexity δ_X and let C be a convex, bounded and closed subset of X. If $T : C \to C$ is uniformly Lipschitzian with constant k and k is less than the (unique) solution of the equation*

$$h(1 - \delta_X(1/h)) = 1,$$

then T has a fixed point in C.

For $X = \ell^2$ the solution of the equation in Theorem 1.2 is $\sqrt{5}/2$. We shall show some improvements of this bound. We start with some results in [L].

DEFINITION 1.3. *Let (M, d) be a metric space. We define the Lifshitz character-istic $\kappa(M)$ to be the supremum of all positive real numbers b such that there exists $a > 1$ such that for every x, y in M and $r > 0$ with $r < d(x, y)$ there exists $z \in M$ satisfying $\overline{B}(x, br) \cap \overline{B}(y, ar) \subset \overline{B}(z, r)$.*

It is clear that $\kappa(M) \geq 1$. In the next theorem, we prove a fixed point result when $\kappa(M) > 1$.

THEOREM 1.4. *Let (M, d) be a complete metric space and $T : M \to M$ a uniformly Lipschitzian mapping with constant $k < \kappa(M)$. If there exists $x_0 \in M$ such that the orbit $\{T^n x_0 : n \in \mathbb{N}\}$ is bounded, then T has a fixed point in M.*

Proof. For any $y \in M$ consider

$$R(y) = \inf\{r \geq 0 : \exists x \in M \quad \text{such that} \quad \{T^n x\} \subset \overline{B}(y, r)\}.$$

The existence of a bounded orbit easily implies that $R(y) < +\infty$ for every $y \in M$. Note that a point $y \in M$ is a fixed point if and only if $R(y) = 0$. Indeed, if y is a fixed point, then for every $\varepsilon > 0$ we have $\{T^n y\} \subset \overline{B}(y, \varepsilon)$ which implies $R(y) = 0$, and conversely if $R(y) = 0$, for every $\varepsilon > 0$ there exists $x \in M$ such that $\{T^n x\} \subset \overline{B}(y, \varepsilon)$. Therefore for every positive integer n we have

$$d(Ty, y) \leq d(Ty, T^n x) + d(T^n x, y)$$
$$\leq kd(y, T^{n-1}x) + \varepsilon \leq (1 + k)\varepsilon.$$

Thus $Ty = y$.

We shall find $y \in M$ such that $R(y) = 0$. Since $k < k(M)$, b is chosen such that $k < b < k(M)$, so there is $a > 1$ such that for every x, y in M and $r > 0$ with $r < d(x, y)$, there exists $z \in M$ such that

$$\overline{B}(x, br) \cap \overline{B}(y, ar) \subset \overline{B}(z, r).$$

We choose $\lambda, 0 < \lambda < 1$, such that

$$\gamma = \min\left\{a\lambda, \frac{b\lambda}{k}\right\} > 1.$$

We shall construct a sequence $\{y_p\} \subset M$ such that

$$R(y_{p+1}) \leq \lambda R(y_p) \quad \text{and} \quad d(y_p, y_{p+1}) \leq (\lambda + \gamma)R(y_p)$$

for every $p \in \mathbb{N}$.

An arbitrary y_1 in M is chosen and the construction of $y_1, \ldots, y_p$ is assumed. If $R(y_p) = 0$, we define $y_{p+1} = y_p$. If $R(y_p) > 0$, there is j such that $d(T^j y_p, y_p) > \lambda R(y_p)$ and $x \in M$ such that

$$\{T^n x\} \subset \overline{B}(y_p, \gamma R(y_p)).$$

Let $\hat{x} = T^j x$. Then for every n we have

$$T^n \hat{x} = T^{n+j} x \in \overline{B}(y_p, \gamma R(y_p)) \subset \overline{B}(y_p, a\lambda R(y_p))$$

and thus

$$d(T^n \hat{x}, T^j y_p) = d(T^{n+j} x, T^j y_p) \leq k d(T^n x, y_p)$$
$$\leq k\gamma R(y_p) \leq b\lambda R(y_p).$$

Consequently,

$$\{T^n \hat{x}\} \subseteq \overline{B}(y_p, a\lambda R(y_p)) \cap \overline{B}(T^j y_p, b\lambda R(y_p)) =: D.$$

Since $b < k(M)$, the definition of this constant implies that there exists $w \in M$ such that D is contained in $\overline{B}(w, \lambda R(y_p))$ and $R(w) \leq \lambda R(y_p)$. Let $y_{p+1} = w$, therefore $R(y_{p+1}) \leq \lambda R(y_p)$, and

$$d(y_{p+1}, y_p) \leq d(y_{p+1}, T^n \hat{x}) + d(T^n \hat{x}, y_p)$$
$$\leq \lambda R(y_p) + \gamma R(y_p) = (\lambda + \gamma) R(y_p).$$

Since $\lambda < 1$, we have

$$R(y_p) \leq \lambda^{p-1} R(y_1) \to 0, \quad (p \to \infty).$$

We claim that $\{y_p\}$ is a Cauchy sequence. Indeed, if $p, l \in \mathbb{N}$ we have

$$d(y_p, y_{p+l}) \leq \sum_{i=0}^{l-1} d(y_{p+i}, y_{p+i+1}) \leq \sum_{i=0}^{l-1} (\lambda + \gamma) R(y_{p+i})$$
$$\leq (\lambda + \gamma) \sum_{i=0}^{l-1} \lambda^{p+i-1} R(y_1) \leq (\lambda + \gamma) \frac{\lambda^{p-1}}{1 - \lambda} R(y_1).$$

We denote $y = \lim_{p \to \infty} y_p$. We shall check that $R(y) = 0$. Indeed, for every $\varepsilon > 0$ there exists y_p such that $R(y_p) < \varepsilon/2$ and $d(y_p, y) < \varepsilon/2$. Thus there exists $x \in M$ such that $\{T^n x\} \subseteq \overline{B}(y_p, \varepsilon/2)$ which implies $\{T^n x\} \subseteq \overline{B}(y, \varepsilon)$ and so $R(y) = 0$. $\square$

Remark 1.5. Nothing can be said about the value of $\kappa(M)$ when M is an arbitrary metric space. When M is a Banach space X we denote by $\kappa_0(X)$ the infimum of the numbers $\kappa(C)$ where C is a closed, convex, and bounded subset of X. The following improved version for Banach spaces of Theorem 1.4 can be found in [Do6]: Let X be a Banach space, C a closed convex bounded subset of X and $T : C \to C$ a k-uniformly Lipschitzian mapping. If

$$k < \frac{1 + \sqrt{1 + 4N(X)(\kappa_0(X) - 1)}}{2}$$

then T has a fixed point.

2. Connections between the Lifshitz characteristic and certain geometric coefficients

In this section we shall prove some connections between the coefficient $\kappa_0(X)$ and other geometric coefficients. The following theorem [DT] is, along with Theorem 1.4, an improved version of Theorem 1.2.

THEOREM 2.1. *Let X be a Banach space and h a solution of the equation*

$$h(1 - \delta_X(1/h)) = 1.$$

Then $h \leq \kappa_0(X)$. Furthermore $\varepsilon_0(X) < 1$ if and only if $\kappa_0(X) > 1$.

Proof. We will only prove that $\varepsilon_0(X) \geq 1$ implies $\kappa_0(X) = 1$, because the other statements will be proved (in a more general form) in Theorem 2.5. We assume $\varepsilon_0(X) \geq 1$ and we arbitrarily choose $b > 1$ and $a > 1$. We let $\gamma = \min\{a, b, 2\} > 1$. There exist two norm one elements x, y in X such that $\|x - y\| > 1/\gamma$ and $\|(x + y)/2\| > 1/\gamma$. Since $\|\gamma x\| = \gamma \leq b$ and $\|\gamma x - \gamma(x - y)\| = \gamma\|y\| = \gamma \leq a$, we have that γx belongs to $\overline{B}(0, b) \cap \overline{B}(\gamma(x - y), a)$. Similarly $-\gamma y$ belongs to $\overline{B}(0, b) \cap \overline{B}(\gamma(x - y), a)$. However $\|\gamma x - (-\gamma y)\| = \gamma\|x + y\| > 2$ and hence there is no $z \in X$ such that $\overline{B}(0, b) \cap \overline{B}(\gamma(x - y), a) \subset \overline{B}(z, 1)$. Since 0 and $\gamma(x - y)$ are in $\overline{B}(0, 4)$ and a, b are arbitrary we obtain $\kappa(\overline{B}(0, 4)) = 1$. Thus $\kappa_0(X) = 1$. $\square$

THEOREM 2.2. *Let X be a Banach space. Then $\kappa_0(X) \leq N(X)$. In particular $\kappa_0(X) \leq \sqrt{2}$.*

Proof. Let C be any closed, bounded and convex set in X with $\operatorname{diam}(C) > 0$. We claim that $b \leq \operatorname{diam}(C)/r(C)$ for every $b < \kappa_0(X)$. Indeed, otherwise $\operatorname{diam}(C)/r(C) < b$. Let $a > 1$ be the number corresponding to b in the definition of $\kappa(C)$. Note that a can chosen small enough so that $ba < \kappa(C)$ and a is also suitable for ab in the definition of $\kappa(C)$. We choose $\varepsilon > 0$ such that $(1 + \varepsilon)/(1 - \varepsilon) < a$, and $z \in C$ such that C is contained in $\overline{B}(z, r(C)(1 + \varepsilon)) \subset \overline{B}(z, r(C)a(1 - \varepsilon))$. By definition of $r(C)$ there exists $y \in C$ such that $\|y - z\| > r(C)(1 - \varepsilon)$. Since $\operatorname{diam}(C) < br(C)$ we have $C \subset \overline{B}(y, br(C)) \subset \overline{B}(y, abr(C)(1 - \varepsilon))$. The condition that ab satisfies for $r = r(C)(1 - \varepsilon)$ implies that there exists $u \in C$ such that C is contained in $\overline{B}(u, r(C)(1 - \varepsilon))$. This is a contradiction because $r(C)(1 - \varepsilon)$ is strictly less than the Chebyshev radius of C. $\square$

Remark 2.3. From Theorem 2.2 it is clear that X is reflexive and has uniform normal structure if $\kappa_0(X) > 1$. Note that the condition $N(X) > 1$ assures the f.p.p. for nonexpansive mapping. The condition $\kappa_0(X) > 1$ assures the f.p.p. for k-uniformly Lipschitzian mappings when $k < \kappa_0(X)$. We shall show in the following example that both conditions can be different.

Example 1: Let E_β, $1 \leq \beta \leq \sqrt{2}$, be the space ℓ^2 renormed by

$$\|x\|_\beta = \max\left\{\|x\|_2, \beta\|x\|_\infty\right\}.$$

It is easy to check that $d(E_\beta, \ell^2) = \beta$. Thus from Theorems VI.2.8 and VI.6.3 we have $N(E_\beta) \geq \sqrt{2}/\beta$. Furthermore, the set $A = \overline{co}(\{e_n : n \in \mathbb{N}\})$ satisfies $\operatorname{diam}(A) = \sqrt{2}$ and $r(A) = \beta$ because $\|x - e_{n+1}\|_\beta = \beta$ if $x \in \operatorname{co}(\{e_1, \ldots, e_n\})$. Thus $N(E_\beta) = \sqrt{2}/\beta$. However we shall prove that $\kappa(E_\beta) = 1$ if $\beta \geq \sqrt{5}/2$. (If $\sqrt{5}/2 > \beta$, it can be proved [Do 6] that $\kappa_0(E_\beta) = (1 + 1/\beta^2 - 2\beta^{-2}\sqrt{\beta^2 - 1})^{1/2}$). Assume $\sqrt{2} > \beta \geq \sqrt{5}/2$, $b > 1$ and $1 < a \leq b$. Since $0 < \beta^2 - 1 < 1$ there is $t \in \mathbb{R}$ such that

$$\frac{a}{\beta} > t > \max\left\{\frac{a\sqrt{\beta^2 - 1}}{\beta}, \frac{1}{\beta}\right\}.$$

Since $4(\beta^2 - 1) \geq 1$ we have

$$4\beta^2(a^2 - t^2) > 4\beta^2\left(a^2 - \frac{a^2}{\beta^2}\right) = 4a^2(\beta^2 - 1) \geq a^2 > 1.$$

Hence, there exists $s \in \mathbb{R}$ such that $2\sqrt{a^2 - t^2} > s > 1/\beta$. Denote $d = \sqrt{a^2 - t^2}$ and consider the vectors $x = (-d, 0, \ldots)$, $y = (s - d, 0, \ldots)$, $u = (0, t, 0, \ldots)$ and $v = -u$. Then, we have

$$\|x - y\|_\beta = \max\{s, \beta s\} = \beta s > 1.$$

The condition $t > a\sqrt{\beta^2 - 1}/\beta$ implies

$$\beta d > \sqrt{\beta^2\left(a^2 - \frac{a^2(\beta^2 - 1)}{\beta^2}\right)} = a.$$

Furthermore we have $|d - s| \leq d$ because $0 < s < 2d$. So,

$$\|u - x\|_\beta = \|v - x\|_\beta = \max\{a, \beta d, \beta t\} = a \leq b$$

and

$$\|u - y\|_\beta = \|v - y\|_\beta = \max\left\{\sqrt{t^2 + (s - d)^2}, \beta t, \beta|d - s|\right\} \leq a.$$

Therefore $u, v \in \overline{B}(x, b) \cap \overline{B}(y, a)$ and this set is not contained in $\overline{B}(z, 1)$ for any $z \in E_\beta$ because $\|u - v\|_\beta = \max\{2t, 2\beta t\} = 2\beta t > 2$.

In Theorem VI.2.2 we have proved that for any Banach space X one has $N(X) \geq (1 - \delta(1))^{-1}$. We shall prove [L] that the same lower bound holds for $\kappa_0(X)$. Thus we have the relationship $(1 - \delta(1))^{-1} \leq \kappa_0(X) \leq N(X)$ for any Banach space.

LEMMA 2.4. *Let X be a Banach space. Then*

$$\kappa_0(X) \geq \sup\{b > 0 : \text{ there exists } a > 1 \text{ such that for all } y \in X$$

with $\|y\| > 1$, there exists $t \in [0, 1]$ such that $\overline{B}(0, b) \cap \overline{B}(y, a) \subset \overline{B}(ty, 1)\}$.

Proof. We denote by h the number defined on the right-hand side of the above inequality. If $b < h$, there exists $a > 1$ such that for all $y \in X$ with $\|y\| > 1$, there exists $t \in [0,1]$ with

$$\overline{B}(0,b) \cap \overline{B}(y,a) \subset \overline{B}(ty,1).$$

Let C be a closed, convex, and bounded subset of X with $\text{diam}(C) > 0$. For any $x, y \in C$ and $r > 0$, if $\|x - y\| > r$, we have $\|(x - y)/r\| > 1$. Hence there is $t' \in [0,1]$ such that

$$\overline{B}(0,b) \cap \overline{B}((y - x)/r, a) \subset \overline{B}(t'(y - x)/r, 1).$$

Therefore

$$\overline{B}(x, br) \cap \overline{B}(y, ar) = x + r\left(\overline{B}(0,b) \cap \overline{B}((y - x)/r, a)\right)$$
$$\subset x + r\overline{B}(t'(y - x)/r, 1) = \overline{B}(x + t'(y - x), r).$$

Since $x + t'(y - x)$ belongs to C we obtain $\kappa(C) \geq b$ and so $\kappa_0(X) \geq b$. $\qquad \square$

THEOREM 2.5. *Let X be a Banach space. Then*

$$\kappa_0(X) \geq \frac{1}{1 - \delta_X(1)}.$$

Proof. If $\delta_X(1) = 0$, there is nothing to prove. Thus suppose $\delta_X(1) > 0$ and let $1 < b < (1 - \delta_X(1))^{-1}$. Since $\delta_X(\cdot)$ is continuous on $[0,2)$, there is $c \in (1/b, 1)$ such that $b < (1 - \delta_X(c))^{-1}$. Let $a = \min\{bc, 1 + b(1 - c)\} > 1$. We claim that for any $y \in X$ with $\|y\| > 1$, there is $t \in [0,1]$ such that

$$\overline{B}(0,b) \cap \overline{B}(y,a) \subset \overline{B}(ty, 1).$$

If $\|y\| \geq b$, then for any $x \in \overline{B}(0,b) \cap \overline{B}(y,a)$, we have $\|(x - y)/b\| \leq a/b \leq c < 1$, $\|x/b\| \leq 1$ and $\|x/b - (x - y)/b\| = \|y/b\| \geq 1$. Hence

$$\delta_X(1) \leq 1 - \frac{1}{2}\left\|\frac{x}{b} + \frac{x - y}{b}\right\| = 1 - \frac{1}{b}\left\|x - \frac{y}{2}\right\|,$$

that is $\|x - y/2\| \leq b(1 - \delta_X(1)) < 1$. Therefore $\overline{B}(0,b) \cap \overline{B}(y,a) \subset \overline{B}(y/2, 1)$.

If $b > \|y\| \geq a$, then for any $x \in \overline{B}(0,b) \cap \overline{B}(y,a)$, we have $\|x/b\| \leq 1$, and

$$\left\|\frac{x}{b} - \frac{y}{\|y\|}\right\| \leq \frac{\|x - y\|}{b} + \left|\frac{1}{b} - \frac{1}{\|y\|}\right|\|y\|$$
$$\leq \frac{a}{b} + \left|\frac{\|y\|}{b} - 1\right| = \frac{a}{b} + 1 - \frac{\|y\|}{b} \leq 1.$$

Since $\|x/b - (x/b - y/\|y\|)\| = 1$ we have

$$\frac{1}{2}\left\| \frac{x}{b} + \left(\frac{x}{b} - \frac{y}{\|y\|} \right) \right\| \le 1 - \delta_X(1),$$

that is,

$$\left\| x - \frac{by}{2\|y\|} \right\| \le b(1 - \delta_X(1)) < 1.$$

Therefore $\overline{B}(0, b) \cap \overline{B}(y, a)$ is contained in $\overline{B}(by/2\|y\|, 1)$.

Finally when $a > \|y\| > 1$, we can set $\lambda = c/\|y\|$. Then for any $x \in \overline{B}(0, b) \cap \overline{B}(y, a)$, we have $\|x/b\| \le 1$ and

$$\left\| \frac{x}{b} - \lambda y \right\| \le \frac{\|x - y\|}{b} + \left| \frac{1}{b} - \lambda \right| \|y\| \le \frac{a}{b} + \left| \frac{\|y\|}{b} - c \right|$$

$$\le \frac{a}{b} + c - \frac{1}{b} \le \frac{1 + b(1 - c)}{b} + c - \frac{1}{b} = 1.$$

Since $\|x/b - (x/b - \lambda y)\| = \lambda\|y\| = c$, we have $\|x/b + (x/b - \lambda y)\|/2 \le 1 - \delta_X(c)$, that is, $\|x - (b\lambda/2)y\| \le b(1 - \delta_X(c)) < 1$. Therefore $\overline{B}(0, b) \cap \overline{B}(y, a)$ is contained in $\overline{B}((b\lambda/2)y, 1)$. From Lemma 2.4 we obtain that $\kappa_0(X) \ge (1 - \delta_X(1))^{-1}$ in every case. $\qquad \square$

Remark 2.6. Let h be a solution of the equation $h(1 - \delta_X(1/h)) = 1$. Then $1 - \delta_X(1) \le 1 - \delta_X(1/h) = 1/h$ because δ_X is increasing. Thus $(1 - \delta_X(1))^{-1} \ge h$. For $X = \ell^2$ we have $(1 - \delta_X(1))^{-1} = 2/\sqrt{3}$ which is strictly greater than $\sqrt{5}/2$, the bound obtained from Theorem 1.2. However, as we will soon see, neither bounds are sharp.

Although some lower bounds for $\kappa_0(X)$ in special spaces are known (see [AX], [Gr2] and [WZ]), the exact value of $\kappa_0(X)$ is unknown for almost all Banach spaces. In the following theorem we compute $\kappa_0(X)$ when X is a Hilbert space. The value of $\kappa_0(X)$ in some other Banach spaces can be found in [Do6]. The value of $\kappa_0(\ell^p)$ still remains unknown.

THEOREM 2.7. *Let H be an infinite dimensional Hilbert space. Then $k_0(H) = \sqrt{2}$.*

Proof. From Theorem 2.2 we know that $\kappa_0(H) \le \sqrt{2}$. We shall prove the converse inequality. It is easy to prove the equality

$$\|x - ty\|^2 = (1 - t)\|x\|^2 + t\|x - y\|^2 + (t^2 - t)\|y\|^2$$

for any real number t and $x, y \in H$. Let $0 < b < \sqrt{2}$ and let t be chosen such that $b^2/2 < t < 1$. If

$$a = \sqrt{\frac{2 - b^2(1 - t)}{2t}}$$

then, since $b^2 < 2$ we have $a^2 > (2 - 2(1-t))/(2t) = 1$ and $a > 1$. For any $y \in H$ with $\|y\| > 1$, let $x \in \overline{B}(0,b) \cap \overline{B}(y,a)$. Then we have

$$\|x - ty\|^2 = (1-t)\|x\|^2 + t\|x - y\|^2 + (t^2 - t)\|y\|^2 \le (1-t)b^2 + ta^2 - (t - t^2)$$

$$= (1-t)b^2 + \frac{2 - b^2(1-t)}{2} - t(1-t) = 1 + \left(\frac{b^2}{2} - t\right)(1-t) < 1.$$

Thus $\overline{B}(0,b) \cap \overline{B}(y,a)$ is contained in $\overline{B}(ty,1)$ and from Lemma 2.4, $\kappa_0(H) \ge \sqrt{2}$.

□

It is unknown if $\sqrt{2}$ is the best upper bound for k to assure the existence of a fixed point for k-uniformly Lipschitzian mappings in Hilbert spaces. The following example [L] shows that k must be less than $\pi/2$.

Example 2: Let $X = \ell^2$, and consider the subsets

$$B_1^+ = \{x \in \ell^2 : \|x\| \le 1, \quad x_1 = 0 \quad \text{and} \quad x_j \ge 0, \forall j \ge 2\},$$
$$S^+ = \{x \in \ell^2 : \|x\| = 1, \ x_j \ge 0 \ \forall j \ge 1\},$$
$$S_1^+ = S^+ \cap B_1^+,$$

and let $e_1 = (1,0,0,\dots)$. Define the mapping $R : B_1^+ \to S^+$ as

$$R(x) = \cos\left(\frac{\pi}{2}\|x\|\right) e_1 + \sin\left(\frac{\pi}{2}\|x\|\right) \frac{x}{\|x\|}$$

if $x \ne 0$ and $R(0) = e_1$. This mapping has the following properties:

(i) It is the identity mapping on S_1^+.
(ii) It is $(\pi/2)$-Lipschitzian. Furthermore $(\pi/2)$ is its best Lipschitz constant.

Indeed, for $x \ne y$ in B_1^+ consider the function

$$p(t) = R((1-t)x + ty), \quad t \ge 0.$$

Since $\|p'(t)\| \le (\pi/2)\|x - y\|$ for every $t \ge 0$, we have

$$\|R(x) - R(y)\| \le \int_0^1 \|p'(t)\| dt \le \frac{\pi}{2}\|x - y\|.$$

On the other hand if $x, y \in B_1^+$ with $\|x\| = \|y\| = r$ we have

$$\|R(x) - R(y)\| = \frac{\|x - y\|}{r} \sin\left(\frac{\pi}{2}r\right)$$

and $\sin(r\pi/2)/r$ converges to $\pi/2$ as $r \to 0$.

Let us construct a fixed point free mapping. Consider the left shift operator $Q : S^+ \to S_1^+$ defined by $Q(x_1, x_2, \dots) = (0, x_1, x_2, \dots)$, and the composition $Q \circ R$ which gives $T = Q \circ R : B_1^+ \to S_1^+$. Since by (i) $T^n = Q^n \circ R$ and Q is an isometric mapping, we know that T is $(\pi/2)$-uniformly Lipschitzian. Furthermore T is a fixed point free mapping on B_1^+.

Although the bound $\pi/2$ could seem unusual in this kind of results, in [T] it is proved that every k-uniformly Lipschitzian expansive mapping from a bounded closed convex subset C of a Hilbert space into C with $k < \pi/2$ has a fixed point. (We recall that T is called expansive if, $d(Tx, Ty) \geq d(x, y)$ for every x, y.)

3. The normal structure coefficient and fixed points

We have studied fixed point theorems for uniformly Lipschitzian mappings considering the Clarkson modulus and the Lifshitz characteristic. We shall now study the same problem in connection with the normal structure coefficient following [CM]. We shall use the notions of asymptotic radius and asymptotic centre defined in VI.1.

LEMMA 3.1. *Let X be a Banach space with $N(X) > 1$. Then for any bounded sequence $\{x_n\}$ in X, there exists $z \in \overline{co}(\{x_n\})$ such that*

(i) $r_a(\{x_n\}, z) \leq N(X)^{-1} \operatorname{diam}_a(\{x_n\})$.

(ii) *For every $y \in X$ we have $\|z - y\| \leq r_a(\{x_n\}, y)$.*

Proof. For any $k \geq 1$, consider the set

$$A_k = \overline{co}(\{x_n\}_{n \geq k}) \quad \text{and} \quad A = \bigcap_{k=1}^{\infty} A_k.$$

Note that condition (ii) is satisfied for every $z \in A$ because for every $y \in X$ one has

$$\|z - y\| \leq \lim_{k \to \infty} r(A_k, y) = \lim_{k \to \infty} \sup_{n \geq k} \|x_n - y\| = \limsup_{n \to \infty} \|x_n - y\| = r_a(\{x_n\}, y).$$

Since $N(X) > 1$, X is reflexive which implies that A_k is weakly compact. Therefore $A \neq \emptyset$. We shall find $z \in A$ which satisfies (i). The weak lower semicontinuity of $r_a(\{x_n\}, x)$ (see the proof of Lemma VI.3.8) implies that $Z_a(\{x_n\}, A_k)$ and $Z_a(\{x_n\}, A)$ are nonempty. Then for every k consider

$$z_k \in Z_a(\{x_n\}, A_k)$$

which gives a sequence $\{z_k\}$ in the weakly compact A_1. A weakly convergent subsequence $\{z_{n_k}\} \to z$ is taken. Since

$$z \in \overline{co}(\{z_{n_k}\}_{k \geq j}) \subset A_j$$

for every j, we have $z \in A$. We claim that z satisfies (ii). Indeed, we have $z_k \in Z_a(\{x_n\}, A_k)$ and $r_a(\{x_n\}, z_k)$ is a non-decreasing sequence bounded from above by $r_a(\{x_n\}, A)$. Again using the weak lower semicontinuity of the function $r_a(\{x_n\}, x)$ we have

$$\lim_{k \to \infty} r_a(\{x_n\}, z_k) = \lim_{j \to \infty} r_a(\{x_n\}, z_{k_j})$$
$$\geq r_a(\{x_n\}, z) \geq r_a(\{x_n\}, A).$$

Hence

$$\lim_{k \to \infty} r_a(\{x_n\}, z_k) = r_a(\{x_n\}, z) = r_a(\{x_n\}, A).$$

Furthermore, every k satisfies

$$r_a(\{x_n\}, z_k) = r_a(\{x_n\}_{n \geq k}, z_k) = r_a(\{x_n\}_{n \geq k}, A_k)$$
$$\leq r(\{x_n\}_{n \geq k}, A_k) \leq N(X)^{-1} \operatorname{diam}(\{x_n\}_{n \geq k}).$$

Thus, we finally obtain

$$r_a(\{x_n\}, z) \leq N(X)^{-1} \lim_{k \to \infty} \operatorname{diam}(\{x_n\}_{n \geq k}) = N(X)^{-1} \operatorname{diam}_a(\{x_n\}).$$

$\square$

THEOREM 3.2. *Let X be a Banach space, C a closed, convex, bounded subset of X and $T : C \to C$ a k-uniformly Lipschitzian mapping with $k < \sqrt{N(X)}$. Then T has a fixed point.*

Proof. Let x be an arbitrary point in C and consider the sequence $\{T^n x\}$ and the point $z = z(x)$ obtained by application of Lemma 3.1 to this sequence. By (i) in Lemma 3.1 we have

$$r_a(\{T^n x\}, z) \leq N(X)^{-1} \operatorname{diam}_a(\{T^n x\}) \leq N(X)^{-1} \sup_{n \geq m \geq 0} \|T^n x - T^m x\|$$
$$\leq k N(X)^{-1} \sup_{i \geq 1} \|T^i x - x\| = k N(X)^{-1} r(x),$$

where $r(x) = \sup_{i \geq 1} \|T^i x - x\|$. On the other hand, for $N > 1$ we have

$$r_a(\{T^n x\}, T^N z) = \limsup_{n \to \infty} \|T^n x - T^N z\| \leq$$
$$\leq k \limsup_{n \to \infty} \|T^{n-N} x - z\| = k r_a(\{T^n x\}, z).$$

These inequalities and Lemma 3.1(ii) imply for every n that

$$\|T^n z - z\| \leq r_a(\{T^m x\}, T^n z)$$
$$\leq k r_a(\{T^n x\}, z) \leq k^2 N(X)^{-1} r(x).$$

Thus
$$r(z) \leq k^2 N(X)^{-1} r(x) = \eta r(x)$$
where $\eta = k^2 N(X)^{-1}$. Now consider the sequence $\{x_n\}$ defined in the following way: $x_1 \in C$ is chosen arbitrary and $x_{n+1} = z(x_n)$. Note that $\{x_n\}$ is a Cauchy sequence. Indeed,

$$\begin{aligned} \|x_{n+1} - x_n\| &\leq \|x_{n+1} - T^j x_n\| + \|T^j x_n - x_n\| \\ &\leq \|x_{n+1} - T^j x_n\| + r(x_n). \end{aligned}$$

Taking the upper limit as $j \to \infty$ we obtain

$$\begin{aligned} \|x_{n+1} - x_n\| &\leq r_a(\{T^j x_n\}, z(x_n)) + r(x_n) \\ &\leq (1 + kN(X)^{-1}) r(x_n) \end{aligned}$$

which implies
$$\|x_{n+1} - x_n\| \leq (1 + kN(X)^{-1}) \eta^n r(x_1).$$

This inequality proves that $\{x_n\}$ is a Cauchy sequence because $\eta < 1$. We denote by $y \in C$ the limit of $\{x_n\}$. Then y is a fixed point of T. Indeed,

$$\begin{aligned} \|y - Ty\| &\leq \|y - x_n\| + \|x_n - Tx_n\| + \|Tx_n - Ty\| \\ &\leq (1 + k)\|x_n - y\| + r(x_n) \to 0 \quad (n \to \infty). \end{aligned}$$

$$\square$$

Remark 3.3. In Example 1 we have proved that $\kappa_0(E_\beta) = 1$ if $\beta \geq \sqrt{5}/2$ and $N(E_\beta) = \sqrt{2}/\beta$. Thus, Theorem 3.2 assures the existence of fixed point for uniformly Lipschitzian mapping if $k < (\sqrt{2}/\beta)^{1/2}$. Note that, in this case, Theorem 1.4 cannot be used. However for a Hilbert space Theorem 1.4 assures the existence of fixed point if $k < \sqrt{2}$ and Theorem 3.2 can only be used when $k < 2^{1/4}$.

Chapter IX

Asymptotically Regular Mappings

We shall study in this chapter the existence of fixed points for a different class of mappings, called asymptotically regular mappings. The concept of asymptotically regular mappings is due to Browder and Petryshyn [BP]. Some fixed point theorems for this class of mappings can be found in [Gr1], [Gr2] and references therein. The fixed point theorems which we shall study are based upon results in [DX]. As we shall see, there is a strong connection between these results and those in Chapter VIII. In particular, in some of them the role of the Clarkson modulus of convexity will be played by the moduli of near uniform convexity. In Section 1 we define a new geometric coefficient in Banach spaces which plays the role of the Lifshitz characteristic for asymptotically regular mappings, and we prove the corresponding version for these mappings of Theorem VIII.1.4. In Section 2 we study some relationships between the new coefficient and either the modulus of NUC or the weakly convergent sequence coefficient. We also find a simpler expression for the new coefficient in Banach spaces with the uniform Opial property. Moreover we prove that, in contrast to the Lifshitz characteristic, the new coefficient is easy to compute in ℓ^p-spaces. We recall that the Lifshitz characteristic is only known in some renorming of Hilbert spaces.

David Hilbert (1862–1943) was born in Wehlan (Germany), near to Königsberg, on 23 June. At that time Otto, his father, was a county judge. Shortly after David's birth, Otto was nominated as judge in Königsberg, moving himself and his family there. In this city Hilbert completed his first studies and subsequently entered the University. Besides Königsberg, he studied in Heidelberg and Berlin and had some of the best mathematicians of the time as lecturers (Weierstrass, Fuchs, Kummer, Kronecker, Weber, etc.). On 11 December 1884 he attained his doctor's degree with a thesis on algebraic invariants directed by Lindemann.

In July 1886 Hilbert was admitted as Privatdozent, and in 1892 he married Käthe Jerosh. In this same period he was appointed Ausserordentlicher Professor succeeding his teacher and friend Adolf Hurwitz.

In the following year, he advanced to a full professorship by replacing Lindemann who left for Berlin to fill a new post.

Finally, in 1895 he was appointed Mathematics Professor in Göttingen. He was to spend the rest of his life in this city, making the mathematics department of his university one of the main mathematical centres in the world at that time.

Throughout his scientific career Hilbert received numerous awards and distinctions, becoming one of the most prestigious mathematicians of his time.

In Section 3 we show that the weakly convergent sequence coefficient can also be used to prove a fixed point theorem for asymptotically regular mappings.

1. A fixed point theorem for asymptotically regular mappings

DEFINITION 1.1. *Let (X, d) be a metric space. A mapping $T : M \to M$ is called asymptotically regular if*

$$\lim_{n \to \infty} d(T^n x, T^{n+1} x) = 0$$

for all $x \in M$.

Example 1: Let $T : [0, 1] \to [0, 1]$ be an arbitrary nonexpansive mapping. It is easy to check that $S = (I + T)/2$ is also nonexpansive. Thus

$$|S^{n+1}x - S^n x| \leq \cdots \leq |S^2 x - Sx| \leq |Sx - x|.$$

Furthermore S is a nondecreasing function. Indeed, if $x \leq y$ and $Sx > Sy$ we have $(x + Tx)/2 > (y + Ty)/2$ which implies $|Tx - Ty| \geq Tx - Ty > y - x = |x - y|$. Thus

$$1 \geq |S^{n+1}x - x| = \sum_{k=1}^{n} |S^{k+1}x - S^k x| \geq n|S^{n+1}x - S^n x|$$

which implies $|S^{n+1}x - S^n x| \leq 1/n$. Then S is asymptotically regular. In fact, it can be proved (see [GK1, Theorem 9.4, page 98]) that if T is a nonexpansive mapping from a bounded convex subset C of a Banach space into C, then $T_\lambda = \lambda I + (1 - \lambda)T$ is asymptotically regular for all $0 < \lambda < 1$.

An extension of the Lifshitz Theorem (Theorem VIII.1.4) to asymptotically regular mappings can be found in [Gr1]. When asymptotically regular mappings are considered we shall prove in Theorem 1.4 that the Lifshitz characteristic can be replaced by another geometric coefficient which can be computed in certain classes of Banach spaces.

Let X be a Banach space and M be a nonempty, bounded, convex subset of X. Suppose τ is a topology on X. We are going to introduce the new coefficient $\kappa_\tau(M)$, the τ-characteristic of M, which plays the role of the Lifshitz characteristic for asymptotically regular mappings.

DEFINITION 1.2.

(i) *A number $b \geq 0$ is said to have the property (P_τ) with respect to M if there exists $a > 1$ such that for all $x, y \in M$ and $r > 0$ with $\|x - y\| \geq r$ and each τ-convergent sequence $\{\xi_n\} \subset M$ for which $\limsup_{n\to\infty} \|\xi_n - x\| \leq ar$ and $\limsup_{n\to\infty} \|\xi_n - y\| \leq br$, there exists some $z \in M$ such that $\liminf_{n\to\infty} \|\xi_n - z\| \leq r$.*

(ii) $\kappa_\tau(M) = \sup\{b > 0 : b \text{ has property } (P_\tau) \text{ with respect to } M\}$.

(iii) $\kappa_\tau(X) = \inf\{\kappa_\tau(M) : M \text{ is as above }\}$.

If τ is the weak topology $\sigma(X, X^*)$ of X, then we write $\kappa_w(M)$ and $\kappa_w(X)$ instead of $\kappa_{\sigma(X,X^*)}(M)$ and $\kappa_{\sigma(X,X^*)}(X)$, respectively.

Remark 1.3. It is easily seen that $\kappa_\tau(M) \geq \kappa(M)$ for every bounded convex subset $M \subset X$.

If S is a mapping from a set C into itself, then we use the symbol $|S|$ to denote the Lipschitz constant of S, that is,

$$|S| = \sup\left\{\frac{\|Sx - Sy\|}{\|x - y\|} : x, y \in C, x \neq y\right\}.$$

Now if T is a mapping on C, we set

$$s(T) = \liminf_{n\to\infty} |T^n|.$$

THEOREM 1.4. *Suppose X is a Banach space and τ is a topology on X. Suppose also that C is a bounded, convex subset of X and $T : C \to C$ is an asymptotically regular mapping. If C is τ-sequentially compact and $s(T) < \kappa_\tau(C)$, then T has a fixed point.*

Proof. A sequence $\{n_k\}$ of positive integers is chosen such that $s(T) = \lim_{k\to\infty} |T^{n_k}|$ and we set $L_k = |T^{n_k}|$. We define a function r on C by

$$r(x) = \inf\left\{d > 0 : \exists y \in C \quad \text{such that} \quad \liminf_{k\to\infty} \|x - T^{n_k}y\| \leq d\right\}.$$

The mapping r is well defined because C is bounded. Since $s(T) < \kappa_\tau(C)$, it is readily seen that there exist positive numbers $\alpha, \mu \in (0,1)$ such that for every $x, y \in C$ and $r > 0$ with $\|x-y\| \geq (1-\mu)r$ and any τ-convergent sequence $\{\xi_n\}$ in C for which $\limsup_{n\to\infty} \|\xi_n - x\| \leq (1+\mu)r$ and $\limsup_{n\to\infty} \|\xi_n - y\| \leq s(T)(1+\mu)r$, there exists some $z \in C$ such that $\liminf_{n \to \infty} \|\xi_n - z\| \le \alpha r$. Without any loss of generality, we may assume that $L_k < (1+\mu)^{\frac{1}{2}} s(T)$ for all $k \ge 1$. From the definition of $r(x)$, there exists certain integer $m \ge 1$ such that $\|x - T^{nm}x\| > r(x)(1 - \mu)$ and a certain $y \in C$ such that

$$\liminf_{n \to \infty} \|x - T^{n_k}y\| \le (1+\mu)^{\frac{1}{2}} r(x). \tag{1}$$

We take a subsequence $\{T^{n_{k'}}y\}$ of $\{T^{n_k}y\}$ such that

$$\liminf_{k \to \infty} \|x - T^{n_k}y\| = \lim_{k' \to \infty} \|x - T^{n_{k'}}y\|.$$

Since C is τ-sequentially compact, we may assume that $\{T^{n_{k'}}y\}$ is τ-convergent. Note that the asymptotic regularity of T over C implies

$$\lim_{k' \to \infty} \|T^{n_{k'}+m}y - T^{n_{k'}}y\| = 0$$

for every fixed integer m. Therefore we derive that

$$\limsup_{k' \to \infty} \|T^{n_{k'}}y - T^{nm}x\| \le L_m \limsup_{k' \to \infty} \|T^{n_{k'}-nm}y - x\|$$

$$= L_m \limsup_{k' \to \infty} \|T^{n_{k'}}y - x\| = L_m \liminf_{k \to \infty} \|T^{n_k}y - x\|$$

$$\le (1+\mu)s(T)r(x). \tag{2}$$

It follows from (1) and (2) that there exists $z = z(x) \in C$ such that

$$\liminf_{k' \to \infty} \|T^{n_{k'}}y - z\| \le \alpha r(x),$$

which implies that

$$r(z) \le \alpha r(x).$$

Also, we have

$$\|z - x\| \le \liminf_{k' \to \infty} \|z - T^{n_{k'}}y\| + \limsup_{k' \to \infty} \|T^{n_{k'}}y - x\|$$

$$\le \alpha r(x) + (1+\mu)r(x) = Ar(x),$$

where $A = 1 + \alpha + \mu$. Proceeding in this way, we obtain a sequence $\{z_n\}$ in C ($z_0 = x$ and $z_k = z(z_{k-1})$) such that

$$r(z_{n+1}) \le \alpha r(z_n) \quad \text{and} \quad \|z_{n+1} - z_n\| \le Ar(z_n). \tag{3}$$

Therefore it can be easily seen from (3) that $\{z_n\}$ is a norm-Cauchy sequence and thus strongly convergent. Let $z_\infty = \lim_{n \to \infty} z_n$. It is readily seen that $r(z_\infty) = 0$, which implies that z_∞ is a fixed point of T. Indeed, if $r(z_\infty) = 0$ for any $\varepsilon > 0$ there exists $y \in C$ such that

$$\liminf_{k \to \infty} \|z_\infty - T^{n_k} y\| \leq \varepsilon.$$

Take a subsequence $\{T^{n_{k''}} y\}$ of $\{T^{n_k} y\}$ such that

$$\liminf_{k \to \infty} \|z_\infty - T^{n_k} y\| = \lim_{k'' \to \infty} \|z_\infty - T^{n_{k''}} y\|.$$

By assumption there exists a positive integer m such that $|T^m| < \infty$. Therefore

$$\limsup_{k'' \to \infty} \|T^{n_{k''} + m} y - z_\infty\| = \lim_{k'' \to \infty} \|T^{n_{k''}} y - z_\infty\| \leq \varepsilon.$$

Thus

$$\|z_\infty - T^m z_\infty\| \leq \limsup_{k'' \to \infty} \left(\|z_\infty - T^{n_{k''} + m} y\| + \|T^{n_{k''} + m} y - T^m z_\infty\| \right)$$

$$\leq \limsup_{k'' \to \infty} \|z_\infty - T^{n_{k''} + m} y\| + |T^m| \lim_{k'' \to \infty} \|T^{n_{k''}} y - z_\infty\|$$

$$= (1 + |T^m|) \liminf_{k \to \infty} \|z_\infty - T^{n_k} y\|$$

$$< (1 + |T^m|)\varepsilon \quad \longrightarrow 0 \quad \text{as} \quad \varepsilon \to 0,$$

yielding $z_\infty = T^m z_\infty$.

It is easily verified that $T^{ms} z_\infty = z_\infty$ for $s = 1, 2, \dots$. Therefore

$$\|T z_\infty - z_\infty\| = \|T^{ms+1} z_\infty - T^{ms} z_\infty\| \to 0$$

as $s \to \infty$. So $T z_\infty = z_\infty$. $\square$

Remark 1.5. An improved version of Theorem 1.4 can be found in [Do6] when τ is the weak topology.

2. Connections between the τ-characteristic and some other geometric coefficients

In order to better study the τ-characteristic we establish an equivalent definition in Banach spaces satisfying the uniform Opial condition.

THEOREM 2.1. *Suppose X is a Banach space satisfying the uniform Opial condition and M a nonempty bounded convex subset of X. Then*

$$\kappa_w(M) = \sup\{b > 0 : \forall z, y \in M, \forall r > 0 \quad \text{with} \quad \|z - y\| \geq r \quad \text{and every}$$

$$\text{sequence} \quad \{\xi_n\} \subset M \quad \text{converging weakly to} \quad z \in M \quad \text{such}$$

$$\text{that} \quad \limsup_{n \to \infty} \|\xi_n - y\| \leq br, \text{ we have} \quad \liminf_{n \to \infty} \|\xi_n - z\| \leq r\}.$$

Proof. We denote $h(M)$ as the quantity defined by the right-hand side of the above equality. First we show that $h(M) \geq \kappa_w(M)$. This will be reached after we verify

$$h(M) \geq b, \tag{4}$$

whenever b is an arbitrary number satisfying $0 < b < \kappa_w(M)$. To show (4), we assume that z, y are in M and $r > 0$ with $\|z - y\| \geq r$ and that $\{\xi_n\} \subset M$ converges weakly to z with $\limsup_{n\to\infty} \|\xi_n - y\| \leq br$. We want to show that $\limsup_{n\to\infty} \|\xi_n - z\| \leq r$. By translation and multiplication we assume that $r = 1$ and $z = 0$. Since $b < \kappa_w(M)$ there exists $a > 1$ with the following property:

For any $u, v \in M$ and $\rho > 0$ satisfying $\|u - v\| \geq \rho$ and any sequence $\{\eta_n\} \subset M$ converging weakly for which $\limsup_{n\to\infty} \|\eta_n - u\| \leq b\rho$ and $\limsup_{n\to\infty} \|\eta_n - v\| \leq a\rho$, there exists $w \in M$ such that $\liminf_{n\to\infty} \|\eta_n - w\| \leq \rho$.

Now we claim that

$$\limsup_{n\to\infty} \|\xi_n\| \leq a. \tag{5}$$

Indeed, we set $\alpha = \limsup_{k\to\infty} \|\xi_n\|$ and choose a subsequence $\{\xi_{n_k}\}$ of $\{\xi_n\}$ such that $\alpha = \lim_{k\to\infty} \|\xi_{n_k}\|$. Suppose (5) fails. Then $\alpha > a$. We choose any $\alpha' \in (\alpha/a, \alpha)$ (clearly, $\alpha' > 1$) and consider the sequence $\{\eta_k\} \subset M$ defined by

$$\eta_k = \frac{1}{\alpha'}\xi_{n_k}.$$

Then $\eta_k \rightharpoonup 0$ and $\limsup_{k\to\infty} \|\eta_k\| = \frac{1}{\alpha'}\limsup_{k\to\infty} \|\xi_{n_k}\| = \frac{\alpha}{\alpha'} < a$. Noting that $\|y\| \leq \liminf_{k\to\infty} \|\xi_n - y\| \leq \limsup_{k\to\infty} \|\xi_n - y\| \leq b$ from the weak lower semicontinuity of the norm of X, we obtain

$$\begin{aligned}
\limsup_{k\to\infty} \|\eta_k - y\| &= \frac{1}{\alpha'}\limsup_{k\to\infty} \|\xi_{n_k} - \alpha'y\| \\
&\leq \frac{1}{\alpha'}(\limsup_{k\to\infty} \|\xi_{n_k} - y\| + (\alpha' - 1)\|y\|) \\
&\leq \frac{1}{\alpha'}(b + (\alpha' - 1)b) = b.
\end{aligned}$$

Using the property that the number a possesses, we can find $z' \in M$ such that $\liminf \|\eta_k - z'\| \leq 1$. Opial's condition then yields

$$\frac{\alpha}{\alpha'} = \liminf_{k\to\infty} \|\eta_k\| \leq \liminf_{k\to\infty} \|\eta_k - z'\| \leq 1,$$

which contradicts the choice of α' and (5) is proved.

Now using (5) and again the property that a has, we obtain $\liminf_{n\to\infty} \|\xi_n - w\| \leq 1$ for some $w \in M$. Opial's condition then implies that $\liminf_{n\to\infty} \|\xi_n\| \leq 1$. This proves (4).

Conversely we next show $h(M) \leq \kappa_w(M)$. Assume that $0 < b < h(M)$ is arbitrary. We choose $0 < d < 1$ close enough to 1 so that

$$b' := \frac{b}{d} < h(M).$$

From the definition of $h(M)$, b' has the following property:

Given $u, v \in M, \rho > 0$ with $\|u - v\| \geq \rho$ and

$\{\eta_n\} \subset M$ for which $\eta_n \rightharpoonup u$ and $\limsup\limits_{n \to \infty} \|\eta_n - v\| \leq b'\rho$, (6)

then $\liminf\limits_{n \to \infty} \|\eta_n - u\| \leq \rho$.

For $1 - d > 0$, using the uniform Opial condition, there exists $c > 0$ such that

$$\liminf_{n \to \infty} \|x_n - x\| \geq 1 + c$$

for all $x \in X$ with $\|x\| \geq 1 - d$ and weakly null sequences $\{x_n\}$ in X with $\liminf_{n \to \infty} \|x_n\| \geq 1$. Now choose any $a \in (1, 1 + c)$. We shall show the following property:

Given any $x, y \in M, r > 0$ with $\|x - y\| > r$ and any

weakly convergent sequence $\{\xi_n\} \subset M$ with $\limsup\limits_{n \to \infty} \|\xi_n - y\| \leq br$

and $\limsup\limits_{n \to \infty} \|\xi_n - x\| \leq ar$, then there exists $z \in M$ (7)

such that $\liminf\limits_{n \to \infty} \|\xi_n - z\| \leq r$.

As in the proof of the first part, we may assume that $r = 1$ and $\xi_n \rightharpoonup 0$. If $\|x\| \leq 1 - d$, then $\|y\| > d$ and $\limsup_{n \to \infty} \|\xi_n - y\| \leq b = b'd$. Hence it follows from (6) that

$$\liminf_{n \to \infty} \|\xi_n\| \leq d < 1$$

and this case is complete. So assume now that $\|x\| > 1 - d$. We then claim that

$$R := \liminf_{n \to \infty} \|\xi_n\| \leq 1, \tag{8}$$

which will conclude the proof. In fact, if $R > 1$ while $\|x\| > 1 - d$, the uniform Opial condition implies that

$$\liminf_{n \to \infty} \|\xi_n - x\| \geq 1 + c,$$

which leads to the contradiction

$$a \geq \limsup_{n \to \infty} \|\xi_n - x\| \geq 1 + c > a,$$

and so (8) is verified. The proof of the theorem is now complete. $\square$

The following result is parallel to Theorem VIII.2.1, replacing the Clarkson modulus of convexity with the modulus $\Delta_{X, \chi}$ of near uniform convexity corresponding to the Hausdorff measure of noncompactness.

THEOREM 2.2. *Let X be a reflexive Banach space with the uniform Opial condition and M be a bounded convex subset of X. Then $\kappa_w(M) \geq h$, where h is the unique solution of the equation*

$$t\left[1 - \Delta_{X,\chi}\left(\frac{1}{t}\right)\right] = 1.$$

Proof. We set

$$h(t) = \frac{1}{t} + \Delta_{X,\chi}\left(\frac{1}{t}\right).$$

Since $\Delta_{X,\chi}$ is nondecreasing and continuous on $[0,1)$ (see Theorem V.1.18), $h(t)$ is strictly decreasing and continuous on $[0,1)$. On the other hand, the reflexivity and the uniform Opial condition of X implies $\lim_{\varepsilon \to 1^-} \Delta_{X,\chi}(\varepsilon) = 1$ (see Theorem V.3.11) and thus $\lim_{t \to 1^+} h(t) = 2$. Furthermore, $\lim_{t \to +\infty} h(t) = 0$. It follows that there exists a unique $h > 1$ such that

$$h\left[1 - \Delta_{X,\chi}\left(\frac{1}{h}\right)\right] = 1.$$

Now assume $b < h$. Then $h(b) > 1$, that is,

$$\Delta_{X,\chi}\left(\frac{1}{b}\right) > 1 - \frac{1}{b}.$$

Let y, z be points in M such that $\|y-z\| \geq r$ and $\{\xi_n\}$ be a sequence weakly convergent to z such that $\limsup_{n\to\infty} \|y-\xi_n\| \leq br$. We shall prove that $\liminf_{n\to\infty} \|\xi_n - z\| \leq r$ which proves $b \leq \kappa_w(M)$ using Theorem 2.1. By translation we assume that $z = 0$. We claim that

$$\chi(\{\xi_n - y\}) \leq r.$$

Suppose, by contradiction, that $\chi(\{\xi_n - y\}) > r$. Then by considering the sequence $\{(\xi_n - y)/br\}$ which converges weakly to $-y/br$ and using Theorem V.1.11, we obtain

$$1 - \left\|-\frac{y}{br}\right\| \geq \Delta_{X,\chi}\left(\frac{1}{b}\right) > 1 - \frac{1}{b},$$

yielding $\|y\| < r$, a contradiction to the fact $\|y\| \geq r$. Thus $\chi(\{\xi_n\}) \leq r$. The definition of χ implies that for every $\delta > 0$ there exists $z \in X$ such that $\|\xi_n - z\| < r+\delta$ for infinitely many integers n. The Opial condition then implies $\liminf_{n\to\infty} \|\xi_n\| \leq \liminf_{n\to\infty} \|\xi_n - z\| \leq r + \delta$. Since δ is arbitrary, we obtain $\liminf_{n\to\infty} \|\xi_n\| \leq r$. This completes the proof. □

In the same way as the modulus $\Delta_{X,\chi}$ can replace the Clarkson modulus in Theorem VIII.2.1 when κ_w is considered, we shall now show that the modulus $\Delta_{X,\beta}$, corresponding to the separation measure of noncompactness, can also replace the Clarkson modulus in Theorem VIII.2.5 for this new setting.

THEOREM 2.3. *Let X be a Banach space with the uniform Opial condition. Then*

$$\kappa_\omega(X) \geq \frac{1}{1 - \Delta_{X,\beta}(1^-)}.$$

Proof. Let r be a positive real number and $\{x_n\}$ a sequence weakly convergent to z such that $\limsup_{n\to\infty} \|x_n - y\| \leq br$ where $\|y - z\| \geq r$ and $b < (1 - \Delta_{X,\beta}(1^-))^{-1}$. By translation we assume $z = 0$ and consider a subsequence $\{y_n\}$ such that $\liminf_{n\to\infty} \|x_n\| = \lim_{n\to\infty} \|y_n\|$ and $\lim_{n,m\,;n\neq m} \|y_n - y_m\| = lr$. Therefore $\beta(\{y_n\}) = \mathrm{diam}_a(\{y_n\}) = lr$. We claim $l \leq b$. Indeed, otherwise consider the sequence $\{(y_n - y)/(b+\eta)r\}$ for some η, $0 < \eta < (1 - \Delta_{X,\beta}(1^-))^{-1} - b$. Since $\beta(\{(y_n - y)/(b+\eta)r\}) = l/(b+\eta) > l/b$ and $\{(y_n - y)/(b+\eta)r\}$ converges weakly to $-y/(b+\eta)r$ we have from Theorem V.1.11

$$\frac{\|y\|}{(b+\eta)r} \leq 1 - \Delta_{X,\beta}\left(\frac{l}{b}\right).$$

So

$$\|y\| \leq (b+\eta)r\left(1 - \Delta_{X,\beta}\left(\frac{l}{b}\right)\right) \leq (b+\eta)r(1 - \Delta_{X,\beta}(1^-)) < r$$

which is a contradiction. Thus $l \leq b$ and we deduce, using Lemma VI.3.8 (b) and Theorem VI.4.4, that

$$\frac{lr}{\lim_{n\to\infty} \|y_n\|} \geq WCS(X) \geq \frac{1}{1 - \Delta_{X,\beta}(1^-)}$$

which implies

$$\liminf_{n\to\infty} \|x_n\| = \lim_{n\to\infty} \|y_n\| \leq lr(1 - \Delta_{X,\beta}(1^-)) < r.$$

Using Theorem 2.1 we obtain the required inequality. $\qquad\square$

Remark 2.4. For $X = \ell^p, 1 < p < +\infty$, since $\Delta_{\ell^p,\chi}(\varepsilon) = 1 - (1 - \varepsilon^p)^{\frac{1}{p}}$ (see Remark V.1.17) we deduce that the solution h of the equation $t(1 - \Delta_{\ell^p,\chi}(1/t)) = 1$ is $2^{\frac{1}{p}}$. On the other hand, since $\Delta_{\ell^p,\beta}(\varepsilon) = 1 - \left(1 - \frac{\varepsilon^p}{2}\right)^{\frac{1}{p}}$ (see Theorem V.1.16) we also have the value $2^{\frac{1}{p}}$ for the bound $(1 - \Delta_{\ell^p,\beta}(1))^{-1}$ given in Theorem 2.3.

We shall now prove that the coefficient $\kappa_\omega(X)$ is less than $WCS(X)$. We need two lemmas ([Re, Do6]).

LEMMA 2.5. *Let $\{x_n\}$ be a bounded sequence contained in a separable Banach space X. Then a subsequence $\{y_n\}$ of $\{x_n\}$ exists such that $\lim_{n\to\infty} \|y_n - z\|$ exists for all $z \in X$.*

Proof. Let $\{z_n\}$ be dense in X. From a diagonal argument we obtain a subsequence $\{y_n\}$ of $\{x_n\}$ such that $\lim_{n \to \infty} \|y_n - z_m\|$ exists for all m.

Let $z \in X$. We shall prove $\|y_n - z\|$ is a Cauchy sequence. Indeed, for an arbitrary $\varepsilon > 0$ choose z_i such that $\|z_i - z\| < \varepsilon/3$ and $N \in \mathbb{N}$ large enough so that $\|\|y_n - z_i\| - \|y_m - z_i\|\| < \varepsilon/3$ if $m, n \geq N$. Therefore

$$\|\|y_n - z\| - \|y_m - z\|\| \leq \|\|y_n - z_i\| - \|y_m - z_i\|\| + 2\|z - z_i\| < \varepsilon$$

for every $n, m \in \mathbb{N}$. Thus $\{\|y_n - z\|\}$ is a convergent sequence. $\square$

LEMMA 2.6. *Let X be a Banach space without the Schur property. Then*

$$WCS(X) = \inf \left\{ \frac{\lim_{n,m \,; n \neq m} \|x_n - x_m\|}{r_a(x_n)} \right\}$$

where the infimum is taken over all weakly convergent sequences which are not convergent such that $\lim_{n,m \,; n \neq m} \|x_n - x_m\|$ exists and $\lim_{n \to \infty} \|x_n - z\|$ exists for every $z \in \overline{co}(\{x_n\})$.

Proof. Denote by a the value of the right hand side of the above equality. Since $\lim_{n,m \,; n \neq m} \|x_n - x_m\| = \text{diam}_a(\{x_n\})$ if this limit exists, it is clear that $a \geq WCS(X)$. On the other hand, let $\{x_n\}$ be a weakly null sequence such that $\lim_{n,m \,; n \neq m} \|x_n - x_m\|$ exists and $\lim_{n \to \infty} \|x_n - z\|$ exists for every $z \in \overline{co}(\{x_n\})$. Using an argument as in the proof of Lemma VI.3.8 we can prove

$$a \lim_{n \to \infty} \|x_n\| \leq \lim_{n,m \,; n \neq m} \|x_n - x_m\|.$$

To this end, we let $A_k = \overline{co}(\{x_n\}_{n \geq k})$. The weak convergence to zero of $\{x_n\}$ implies that $\bigcap_{k=1}^{\infty} A_k = \{0\}$. Since the function $\Phi(x) = \lim_{n \to \infty} \|x_n - x\|$ is weakly lower semicontinuous (see Lemma VI.3.7) and A_k is weakly compact, Φ attains a minimum at a point z_k in A_k. Since 0 is the unique point which can be weakly adherent to $\{z_k\}$ we infer that $\{z_k\}$ is weakly null. Furthermore it is clear that $\{\Phi(z_k)\}$ is a nondecreasing sequence which is bounded by $\Phi(0)$. Thus $\lim_{k \to \infty} \Phi(z_k) \leq \Phi(0)$. Since the lower semicontinuity of Φ implies $\lim_{k \to \infty} \Phi(z_k) \geq \Phi(0)$ we have $\lim_{k \to \infty} \Phi(z_k) = \Phi(0)$. Since $\Phi(z_k) = r_a(\{x_n\}_{n \geq k})$ we obtain

$$\Phi(z_k)a \leq \lim_{n,m \,; n \neq m} \|x_n - x_m\|.$$

Taking limits as $k \to \infty$ we have

$$a \lim_{n \to \infty} \|x_n\| \leq \lim_{n,m \,; n \neq m} \|x_n - x_m\|.$$

Thus

$$a \leq \left\{ \frac{\lim_{n,m \,; n \neq m} \|x_n - x_m\|}{\lim_{n \to \infty} \|x_n\|} \right\}$$

where the infimum is taken over all weakly null sequences such that $\lim_{n,m \,; n \neq m} \|x_n - x_m\|$ exists and $\lim_{n \to \infty} \|x_n - z\|$ exists for every $z \in \overline{co}(\{x_n\})$. Using Lemma VI.3.8 and Lemma 2.5 it is clear that this infimum is $WCS(X)$. $\square$

THEOREM 2.7. *Let X be a reflexive Banach space. Then $\kappa_w(X) \leq WCS(X)$.*

Proof. Let $\{x_n\}$ be a weakly convergent sequence which is not convergent such that $\lim_{n,m\,;n\neq m} \|x_n - x_m\|$ exists and $\lim_{n\to\infty} \|x_n - z\|$ exists for every $z \in \overline{\text{co}}(\{x_n\})$. Let $l = \lim_{n,m\,;n\neq m} \|x_n - x_m\|$ and $r = r_a(\{x_n\})$. Let b be any number less than $\kappa_w(X)$. We shall prove $b \leq l/r$ which implies the result bearing Lemma 2.6 in mind. Assume, by way of contradiction, that $l < br$, then there exists $k \in \mathbb{N}$ such that $\|x_n - x_m\| < br$ if $m, n \geq k$. Let $a > 1$ be the corresponding number to b in the definition of $\kappa_w(\overline{\text{co}}(\{x_n\}))$. Note that if a is suitable for b in the definition of $\kappa_w(M)$ then every $a' < a$ is also suitable for b and a is suitable for every $b' < b$. We choose $a_1 < a$ suitable for ba in the definition of $\kappa_w(\overline{\text{co}}(\{x_n\}))$. Then a_1 is also suitable for ba_1. Thus we can assume that a is small enough such that $ba < \kappa_w(\overline{\text{co}}(\{x_n\}))$ and that a is also suitable for ba in this definition. We choose $\varepsilon > 0$ such that $a > (1+\varepsilon)/(1-\varepsilon)$ and $z \in \overline{\text{co}}(\{x_n\})$ such that $\lim_{n\to\infty} \|x_n - z\| < r(1+\varepsilon)$. Since $\lim_{n\to\infty} \|x_n - z\| \geq r$ there exists $j \geq k$ such that $\|x_j - z\| \geq r(1 - \varepsilon)$. Therefore we have

$$\limsup_{n\to\infty} \|x_n - x_j\| = \lim_{n\to\infty} \|x_n - x_j\| \leq br < abr(1 - \varepsilon);$$

$$\limsup_{n\to\infty} \|x_n - z\| < r(1 + \varepsilon) < ar(1 - \varepsilon)$$

and $\|x_j - z\| \geq r(1 - \varepsilon)$. Using the condition that ab satisfies there exists $u \in \overline{\text{co}}(\{x_n\})$ such that

$$\lim_{n\to\infty} \|x_n - u\| = \limsup_{n\to\infty} \|x_n - u\| < r(1 - \varepsilon)$$

which is a contradiction because $\lim_{n\to\infty} \|x_n - u\|$ must be bigger or equal to the asymptotic radius r. $\qquad\square$

Remark 2.8. From Theorems 2.3 and 2.7 we obtain the bounds

$$\frac{1}{1 - \Delta_{X,\beta}(1^-)} \leq \kappa_w(X) \leq WCS(X)$$

when X is reflexive and has the uniform Opial condition. It is worth noting the similarity to the bounds

$$\frac{1}{1 - \delta(1)} \leq \kappa_0(X) \leq N(X)$$

which were obtained in Chapter VIII.

COROLLARY 2.9. *If* $1 < p < +\infty$, *then*

$$\kappa_w(\ell^p) = 2^{\frac{1}{p}}.$$

Remark 2.10. Since $\kappa_0(\ell^p) \leq N(\ell^p) = 2^{1-1/p} < 2^{1/p} = \kappa_w(\ell^p)$ we find a Banach space such that $\kappa_0(X)$ is strictly less than $\kappa_w(X)$.

COROLLARY 2.11. *Let C be a bounded closed convex subset of ℓ^p for $1 < p < +\infty$ and $T : C \to C$ be an asymptotically regular mapping such that $s(T) < 2^{\frac{1}{p}}$. Then T has a fixed point.*

We now show an example of a Banach space X with the uniform Opial condition and for which $\kappa_w(X) \neq WCS(X)$.

Example 2: Let $X = \ell^2$ be renormed by

$$\|x\| = \left[|x^1|^p + \left(\sum_{n=2}^{\infty} |x^n|^2 \right)^{\frac{p}{2}} \right]^{\frac{1}{p}}, \quad x = (x^n) \in \ell^2,$$

where $2 < p < +\infty$. Following an argument as in Example VI.3, it is easy to check that $WCS(X) = WCS(\ell^2) = \sqrt{2}$. Since X satisfies the uniform Opial condition we can use Theorem 2.1 to compute $\kappa_w(X)$.

Choosing $y = e_1$, $\xi_n = e_n$ for $n \geq 1$, $b > 2^{\frac{1}{p}}$ and $r = 2^{\frac{1}{p}}/b$, we derive that

$$\|\xi_n - y\| = 2^{\frac{1}{p}} = br \quad \text{and} \quad \|\xi_n\| = 1 > r$$

for every $n \in \mathbb{N}$. Thus, $\kappa_w(X) \leq 2^{\frac{1}{p}}$.

On the other hand, let $\{x_n\}$ be a weakly null sequence in X and $y \in X$ such that $\|y\| \geq 1$. For any vector $x = (x^n) \in X$, denote $\bar{x} = (0, x^2, x^3, \dots)$. Since $\lim_{n \to \infty} x_n^1 = 0$, if $\limsup_{n \to \infty} \|x_n - y\| \leq 2^{1/p}$ we have

$$\limsup_{n \to \infty} \|\bar{x}_n - \bar{y}\|^p = \limsup_{n \to \infty} \|x_n - y\|^p - |y^1|^p \leq 2 - |y^1| \leq 1 + \|\bar{y}\|^p.$$

Using the inequality $1 + t^p \leq (1 + t^2)^{p/2}$ which holds for every $t \in [0,1]$, we obtain

$$\limsup_{n \to \infty} \|\bar{x}^n - \bar{y}\|_2 \leq (1 + \|\bar{y}\|_2)^{1/2}$$

and it is easy to check that this inequality implies $\limsup_{n \to \infty} \|\bar{x}_n\|_2 \leq 1$. Thus $\limsup_{n \to \infty} \|x_n\| = \limsup_{n \to \infty} \|\bar{x}_n\|_2 \leq 1$ and so $\kappa_w(X) \geq 2^{1/p}$.

3. The weakly convergent sequence coefficient and fixed points

We have seen in Theorem VIII.3.2 that a k-uniformly Lipschitzian mapping with $k < \sqrt{N(X)}$ has the fixed point property. We shall now show that $N(X)$ can be replaced by the weakly convergent sequence coefficient $WCS(X)$ when asymptotically regular mappings are considered.

THEOREM 3.1. *Suppose X is a Banach space such that $WCS(X) > 1$, C is a nonempty weakly compact convex subset of X, and $T : C \to C$ is an asymptotically regular mapping such that $s(T) < \sqrt{WCS(X)}$. Then T has a fixed point.*

Proof. Since one can construct a nonempty closed convex separable subset C_0 of C that is invariant under T (that is, $T(C_0) \subset C_0$), we may assume that C itself is separable (see [GK1, page 35]). Following an argument as in the proof of Lemma 2.5, the separability of C makes it possible to select a subsequence $\{n_j\}$ of positive integers such that

$$s(T) = \lim_{j \to \infty} |T^{n_j}| < \sqrt{WCS(X)}$$

and $\{T^{n_j}x\}$ converges weakly for every $x \in C$. We can also assume that

$$\lim_{i,j\,;i \neq j} \|T^{n_i}x - T^{n_j}x\|$$

exists for any $x \in C$.

Now we can construct the sequence $\{x_n\} \subset C$ in the following way: We take $x_0 \in C$ arbitrary and if $x_0, x_1, \ldots, x_{m-1}$ are defined, we define x_m as the weak limit of $\{T^{n_j}x_{m-1}\}$, $(j \to \infty)$: Note that the asymptotic regularity of T on C ensures that $\{T^{n_j+p}x_{m-1}\}$ converges weakly to x_m for every $p \geq 0$.

We now show that $\{x_m\}$ converges strongly to a fixed point of T. To this end, for each integer $m \geq 0$ we write,

$$B_m = \limsup_{j \to \infty} \|T^{n_j}x_m - x_{m+1}\| \quad \text{and} \quad L_m = |T^{n_m}|.$$

Since $\{T^{n_j}x_m - x_{m+1}\}$ is weakly null and using the equivalent definition of $WCS(X)$ in Lemma VI.3.8 we have

$$B_m \leq \frac{\lim_{i,j\,;i\neq j} \|T^{n_i}x_m - T^{n_j}x_m\|}{WCS(X)}.$$

But from the weak lower semicontinuity of the norm of X, it follows that

$$\lim_{i,j\,;i\neq j} \|T^{n_i}x_m - T^{n_j}x_m\| = \limsup_{j \to \infty}\left(\limsup_{i \to \infty} \|T^{n_i}x_m - T^{n_j}x_m\|\right)$$

$$= \limsup_{j \to \infty}\left(\limsup_{i \to \infty} \|T^{n_i+n_j}x_m - T^{n_j}x_m\|\right)$$

$$\leq \left(\limsup_{j \to \infty} L_j\right)\limsup_{i \to \infty} \|T^{n_i}x_m - x_m\|$$

$$\leq s(T)\limsup_{i \to \infty}\left(\liminf_{j \to \infty} \|T^{n_i}x_m - T^{n_j}x_{m-1}\|\right)$$

$$\leq s(T)\left(\limsup_{i \to \infty} L_i\right)\limsup_{j \to \infty} \|x_m - T^{n_j}x_{m-1}\|$$

$$= (s(T))^2 B_{m-1}.$$

Hence

$$B_m \leq \frac{(s(T))^2}{WCS(X)} B_{m-1} = \alpha B_{m-1},$$

where

$$\alpha = \frac{(s(T))^2}{WCS(X)} < 1.$$

Now reapplying the weak lower semicontinuity of the norm of X, we deduce that

$$\|x_m - x_{m+1}\| \leq \limsup_{j \to \infty} \|x_m - T^{n_j} x_m\| + \limsup_{j \to \infty} \|T^{n_j} x_m - x_{m+1}\|$$

$$\leq \limsup_{j \to \infty} \left(\limsup_{i \to \infty} \|T^{n_i} x_{m-1} - T^{n_j} x_m\| \right) + B_m$$

$$\leq s(T) \limsup_{i \to \infty} \|T^{n_i} x_{m-1} - x_m\| + B_m$$

$$= s(T) B_{m-1} + B_m \leq (\sigma(T) B_1 + B_2) \alpha^{m-2},$$

which implies that $\{x_m\}$ is a Cauchy sequence. Let $z = \lim_{m \to \infty} x_m$. Then

$$\|z - T^{n_j} z\| \leq \|z - x_{m+1}\| + \|x_{m+1} - T^{n_j} x_m\| + \|T^{n_j} x_m - T^{n_j} z\|$$

$$\leq \|z - x_{m+1}\| + \|x_{m+1} - T^{n_j} x_m\| + L_j \|x_m - z\|.$$

Taking the limit as $j \to \infty$ yields

$$\limsup_{j \to \infty} \|z - T^{n_j} z\| \leq \|z - x_{m+1}\| + B_m + s(T)\|x_m - z\| \longrightarrow 0$$

as $m \to \infty$. Hence $T^{n_j} z \to z$. From the assumption on $s(T)$ there exists a positive integer m such that T^m is continuous. The asymptotic regularity of T implies

$$T^m(z) = T^m(\lim_{j \to \infty} T^{n_j} z) = \lim_{j \to \infty} T^{n_j + m} z = z.$$

A similar argument as in the proof of Theorem 1.4 lets us conclude that z is a fixed point of T. This completes the proof. □

Remark 3.2. Let X be the Banach space in Example 2. Since $WCS(X) = \sqrt{2}$ and $\kappa_w(X) = 2^{1/p}$, we see that Theorem 3.1 assures the existence of fixed points for k-uniformly Lipschitzian mappings in X if $k < 2^{1/4}$. However Theorem 1.4 can be applied if $k < 2^{1/p}$.

Chapter X

Packing Rates and ϕ-Contractiveness Constants

The main purpose of this chapter is to study relationships between the ϕ-contractiveness constants of an operator when different measures of noncompactness are considered. The first results in this direction were obtained by Nussbaum [N, 1970], Petryshyn [Pe, 1972] and Webb [W1, 1973] for linear mappings.

Some years later more extensive results were obtained in [Do1, 1986] and [Do2, 1988]. In these papers the relationship between set-contractions and ball-contractions in separable Hilbert spaces and certain other classes of spaces is studied.

In fact, in [Do2] it is proved that if X is a separable metric space which has the χ-property (this property is a strong relationship between the separation of the points of a subset of X and the smallest radius of a ball where this subset is contained), then every k-set-contraction is a k-ball-contraction. It is also proved in [Do2] that the spaces ℓ^p, $1 \leq p \leq +\infty$ have the χ-property whereas the Lebesgue spaces $L^p([0,1])$, $p \neq 2, \infty$, do not.

> **Henri Léon Lebesgue** (1875–1941) was born in Beauvais (France) on 28 June and educated at the École Normale Supérieure in Paris between 1894 and 1897. From 1899 to 1902 he worked on his doctoral thesis while he taught mathematics at the lycée in Nancy.
>
> He received his doctorate in 1902 from the Sorbonne and in the same year he was appointed a Lecturer in the faculty of sciences at the University of Rennes. In 1906 he became a Professor at the University of Poitiers, remaining there until 1910 when he was appointed Lecturer in Mathematics at the Sorbonne. In 1921 he took up his final academic post as a Professor of Mathematics at the Collège de France. He died in Paris on 26 July.
>
> Throughout his life, Lebesgue was awarded many honors, including the Prix Houllevique (1912), the Prix Poncelet (1914) and the Prix

Saintour (1917). He was elected to the French Academy of Sciences in 1922 and to the Royal Society in 1934.

Lebesgue obtained significant results in several branches of mathematics such as set theory, calculus of variation and function theory. However, his main contribution was the introduction of a new pattern of integral today known as the Lebesgue integral.

In spite of $L^p(\Omega)$ failing to have the χ-property, it was proved in [AD1, 1991] that the standard relationship between k-set-contractions and k-ball-contractions (nominally: every k-set-contraction is a $2k$-ball-contraction and every k-ball-contraction is a $2k$-set-contraction) can be improved in $L^p(\Omega)$ in the following way: every k-set-contraction in $L^p(\Omega)$ is a $2^{(\frac{|2-p|}{p})}k$-ball-contraction and every k-ball-contraction is a $2^{\max\{2^{\frac{1}{p}-1}, 2^{\frac{-1}{p}}\}}k$-set-contraction. It is clear that the closer p is to 2, the better the improvement is. Conversely, for $p = 1$ no improvement can be obtained from the standard relationship and the situation becomes similar for a very large but finite p.

Moreover, in [AD1] the "packing rate" $\gamma(X)$ of a metric space X was defined. It is a real number in the interval $[1, 2]$ which can be thought as a measure of the relationship between the maximal separation of the points in any subset A of X and the smallest radius of a ball containing A. We can regard X as well "packed" when $\gamma(X)$ is near to 1. This number γ lets us state a relationship between k-set-contractions and k-ball-contractions in X.

The results obtained in [Do1], [Do2] and [AD1] were completed in [ADL, 1990] taking the operators associated with the separation measure of noncompactness into consideration. In [DR2] the coefficient $\gamma(X)$ was computed when X is a direct sum of spaces.

These ideas were generalized in [Ro, 1993] for ϕ-contractive operators for any measure of noncompactness. We are going to follow this general viewpoint throughout the chapter.

1. Comparable measures of noncompactness

We start by giving a very general relationship between ϕ-contractive operators which requires a previous definition.

DEFINITION 1.1. *Let ϕ and λ be two measures of noncompactness in a complete metric space X, and $\mathcal{B}$ the family of bounded sets in X. We shall say that ϕ and λ are comparable measures (or equivalent measures) if the set*

$$\left\{ \frac{\lambda(B)}{\phi(B)} \; : \; B \in \mathcal{B} \, , \, \phi(B) > 0 \right\}$$

is bounded with a positive infimum.

In this case, if we denote by a and b the infimum and the supremum respectively of this set, for any bounded subset B of X we have

$$a\phi(B) \leq \lambda(B) \leq b\phi(B).$$

Moreover, in general, these relations are the best possible between ϕ and λ.

Example 1: From the results obtained in Chapter II, we can conclude that the pairs of measures (β, χ), (β, α) and (χ, α) are comparable measures with constants $a = 1$ and $b = 2$ in all cases.

The first, albeit trivial, result about relationships between ϕ-contractive operators for different measures of noncompactness is the following:

LEMMA 1.2. *Let ϕ and λ be two comparable MNCs defined in a complete metric space X verifying the general relation $a\phi(B) \leq \lambda(B) \leq b\phi(B)$ for any bounded subset B of X. Then, for every mapping $T : D \subset X \to X$, we have the following relationships:*

(a) If T is k-ϕ-contractive, T is (bk/a)-λ-contractive.
(b) If T is k-λ-contractive, T is (bk/a)-ϕ-contractive.

Remark.1.3. If ϕ is an MNC defined in X, for every continuous operator $T : D \subset X \to X$ we define the ϕ-contractiveness constant $\phi(T)$ as

$$\phi(T) = \inf\{k > 0 : T \text{ is } k\text{-}\phi\text{-contractive}\}.$$

For comparable MNCs with $a\phi(A) \leq \lambda(A) \leq b\phi(A)$, the lemma above can be written in the form

$$\frac{a}{b}\phi(T) \leq \lambda(T) \leq \frac{b}{a}\phi(T).$$

Remark 1.4. These relationships cannot be improved in the general class of all Banach spaces for any couple of comparable MNCs.

Example 2: Let $X = \mathcal{C}([0, 1])$ and consider the mapping $T : X \to X$ given by

$$Tx(t) = \begin{cases} \frac{1}{2}x(2t) + \frac{1}{2}x(0) & \text{if } 0 \leq t \leq \frac{1}{2} \\ \frac{1}{2}x(2t - 1) + \frac{1}{2}x(1) & \text{if } \frac{1}{2} < t \leq 1. \end{cases}$$

Then T is a linear and continuous mapping and $\|T\| = 1$ since $\|Tx\| \leq \|x\|$ for all $x \in X$ and $\|Tx_0\| = 1$ for $x_0(t) \equiv 1$. We claim that T is a $(1/2)$-set-contractive operator.

Indeed, let B be a bounded subset of X and $\varepsilon > 0$. We can write $B = \bigcup_{i=1}^{n} A_i$ with $\text{diam}(A_i) \leq \alpha(B) + \varepsilon$ and $A_i \cap A_j = \emptyset$ for all $i \neq j$. Fixing $i \in \{1, 2, \ldots, n\}$ and setting $\varrho = \sup\{|x(0)|, |x(1)| : x \in A_i\}$ we consider intervals J_j of length ε such that $[-\varrho, \varrho] = \bigcup_{j=1}^{p} J_j$. Then $A_i = \bigcup_{j,k=1}^{p} C_{jk}$ with $C_{jk} = \{x \in A_i : x(0) \in$

J_j, $x(1) \in J_k\}$ and $\mathrm{diam}(T(C_{jk})) \leq \alpha(B)/2 + \varepsilon \; \forall j, k = 1, 2, \ldots, p$. Therefore $\alpha(T(A_i)) \leq \alpha(B)/2 + \varepsilon$ for all $i = 1, 2, \ldots, n$. This clearly implies $\alpha(T(B)) \leq (1/2)\alpha(B)$ and so T is a $(1/2)$-set-contractive operator.

Hence T is 1-ball-contractive. Moreover T is not k-ball-contractive for any $k \in [0, 1)$. Indeed, it suffices to notice that the set

$$B = \{x \in X : x(0) = 0, \; x(1) = 1, \; 0 \leq x(t) \leq 1, \; \forall t \in [0, 1]\}$$

verifies $\chi(B) = \chi(T(B)) = 1/2$.

Thus $\alpha(T) = 1/2$ and $\chi(T) = 1$. This relationship is the worst possible between these two MNCs. In the next section we are going to introduce some geometrical coefficients in order to improve the general relation given in Lemma 1.2 in several classes of spaces.

2. Packing rates of a metric space

DEFINITION 2.1. *Let ϕ and λ be two comparable MNCs defined on a metric space X, with a the infimum and b the supremum of the set*

$$\left\{ \frac{\lambda(B)}{\phi(B)} : B \in \mathcal{B}, \; \phi(B) > 0 \right\}.$$

The coefficients $\delta(\lambda, \phi)(X)$ and $\delta'(\lambda, \phi)(X)$ are defined as the supremum and the infimum respectively of the set

$$\left\{ \frac{\lambda(A)}{\phi(A)} : A \subset X, \; A \; \phi\text{-minimal}, \; \phi(A) > 0 \right\}.$$

We denote the coefficient of (λ, ϕ)-packing of X as the number

$$\gamma(\lambda, \phi)(X) = \frac{\delta(\lambda, \phi)(X)}{\delta'(\lambda, \phi)(X)}.$$

When no ambiguity whatsoever about the measures exists, we shall simply write δ, δ' and γ.

Remark 2.2.

(a) Obviously $a \leq \delta' \leq \delta \leq b$ and $1 \leq \gamma(X) \leq \dfrac{b}{a}$ always hold.

(b) We can interpret $\gamma(\lambda, \phi)(X)$ near to 1 as meaning X is (λ, ϕ)-well packed. The most unfavourable case is when $\gamma(\lambda, \phi)(X) = b/a$. Every metric space X is (β, α)-well packed since $\beta(A) = \alpha(A)$ for each α-minimal subset A of X (see Lemma III.2.9). Hence $\delta(\beta, \alpha)(X) = \delta'(\beta, \alpha)(X) = \gamma(\beta, \alpha)(X) = 1$.

Example 3: Let c_0 be the Banach space of all sequences convergent to zero with the supremum norm. Then $\gamma(\chi, \alpha)(c_0) = 2$.

Indeed, for every nonnegative integer n, consider the sequence $\{x_n\}$ given by $x_n = (t_n^m)$ where $t_n^m = -1$ if $m < n$, $t_n^m = 1$ if $m = n$ and $t_n^m = 0$ if $m > n$. Let $A = \{x_n : n \in \mathbb{N}\}$. It is clear that $\|x_i - x_j\| = 2$ for all $i \neq j$. Thus A is α-minimal and $\alpha(A) = 2$. Moreover, since $\|x_i\| = 1$ for all i we have $\chi(A) \leq 1$ and as $\chi(A) \leq \alpha(A) \leq 2\chi(A)$, it follows that $\chi(A) = 1$. Thus $\delta'(\chi, \alpha)(c_0) = 1/2$.

Now let $B = \{y_n : n \in \mathbb{N}\}$ where $y_n = (l_n^m)$ with $l_n^m = 0$ if $m \neq n$ and $l_n^n = 1/2$. Then $\|y_i - y_j\| = 1/2$ for all $i \neq j$ and, therefore B is α-minimal and $\alpha(B) = 1/2$. Furthermore, since $\|y_i\| = 1/2$ for all i, we have $\chi(B) \leq 1/2$. In fact $\chi(B) = 1/2$, because if $r < 1/2$ then $B(x, r) \cap B$ is a finite set for all $x \in c_0$. Thus $\delta(\chi, \alpha)(c_0) = 1$.

Therefore $\gamma(\chi, \alpha)(c_0) = 2$.

Example 4: Let ℓ^∞ be the Banach space of all bounded sequences with the supremum norm. In Example II.2 we proved that $\alpha(A) = 2\chi(A)$ for every bounded subset A of ℓ^∞. Hence $\delta(\chi, \alpha)(\ell^\infty) = 1/2 = \delta'(\chi, \alpha)(\ell^\infty)$ and so $\gamma(\chi, \alpha)(\ell^\infty) = 1$.

In Lemma 1.2 we obtained some trivial relationships between the ϕ-contractiveness constants of a mapping. Now, with the help of the packing rates, we can improve these relations.

THEOREM 2.3. *Let X be a complete metric space with packing rate $\gamma(\lambda, \phi)(X) = \gamma = \delta/\delta'$ with respect to the comparable MNCs ϕ and λ, the latter being minimalizable. Then:*

(a) *If $T : D \subset X \to X$ is k-ϕ-contractive, then T is γk-λ-contractive.*

(b) *If $T : D \subset X \to X$ is k-λ-contractive, then T is $(\delta k/a)$-ϕ-contractive.*

(c) *If X is a Banach space and the measures λ and ϕ are semi-homogeneous in X, the relation (a) cannot be improved, that is:*
   If $\omega < \gamma$, then there are k-ϕ-contractive mappings which are not ωk-λ-contractive mappings.

(d) *If X is a separable Banach space and the measures λ and ϕ are semi-homogeneous in X, the relation (b) cannot be improved, that is:*
   If $\omega < \delta/a$, there are k-λ-contractive mappings which are not ωk-ϕ-contractive mappings.

Proof.

(a) Let A be a bounded and infinite subset of D and let $\varepsilon > 0$. Then there is an infinite subset B of A such that B and $T(B)$ are ϕ and λ-minimal sets and moreover $\lambda(T(A)) \leq \lambda(T(B)) + \varepsilon$. Hence, we have

$$\lambda(T(A)) \leq \lambda(T(B)) + \varepsilon \leq \delta\phi(T(B)) + \varepsilon$$

$$\leq \delta k\phi(B) + \varepsilon \leq \frac{\delta}{\delta'}k\lambda(B) + \varepsilon \leq k\gamma\lambda(A) + \varepsilon.$$

Bearing in mind that $\varepsilon > 0$ was chosen arbitrarily, we obtain $\lambda(T(A)) \leq k\gamma\lambda(A)$, and so the proof of (a) is complete.

(b) For every bounded and infinite subset A of D, we can take B as above and obtain

$$\phi(T(A)) \leq \frac{1}{a}\lambda(T(A)) \leq \frac{1}{a}\left(\lambda(T(B)) + \varepsilon\right)$$
$$\leq \frac{1}{a}(k\lambda(B) + \varepsilon) \leq \frac{1}{a}(k\delta\phi(B) + \varepsilon) \leq \frac{1}{a}k\delta\phi(A) + \frac{\varepsilon}{a}.$$

Since ε was chosen arbitrarily, we obtain

$$\phi(T(A)) \leq \frac{\delta}{a}k\phi(A)$$

and so (b) is proved.

(c) If $\omega < \gamma$, then $\omega\delta' < \delta$ and as δ is a supremum, there exists a ϕ-minimal and nonprecompact set A such that $\omega\delta'\phi(A) < \lambda(A)$. Since δ' is an infimum, there exists B which is ϕ-minimal and nonprecompact verifying that $\phi(A)\lambda(B)\omega < \lambda(A)\phi(B)$. Moreover we can assume that A and B are countable sets

$$A = \{x_n : n \in \mathbb{N}\} \qquad\qquad B = \{y_n : n \in \mathbb{N}\}.$$

On the other hand, since B is nonprecompact we have $\alpha(B) > 0$. Furthermore, for every infinite subset B_0 of B we conclude $\alpha(B_0) > 0$ because otherwise we would obtain $\phi(B_0) = 0$ which contradicts the ϕ-minimality of B.

Let us choose $B_0 \subset B$ α-minimal with $\alpha(B_0) > 0$ and

$$\frac{\alpha(B_0)}{2} < \|x - y\| < \frac{3\alpha(B_0)}{2}$$

for all $x, y \in B_0$, $x \neq y$.

Let us enumerate B_0 as $\{z_n : n \in \mathbb{N}\}$ and note that B_0 is an infinite, ϕ-minimal and discrete set. For $k > 0$ we define the mapping $T : B_0 \to X$ by

$$Tz_n = k\frac{\phi(B)}{\phi(A)} x_n.$$

Obviously T is continuous and by the semi-homogeneity of ϕ and λ, for every infinite subset C of B_0, we obtain

$$\phi(T(C)) \leq \phi(T(B_0)) = k\frac{\phi(B)}{\phi(A)}\phi(A) = k\phi(B) = k\phi(B_0) = k\phi(C)$$

and so T is k-ϕ-contractive. On the other hand,

$$\lambda(T(B_0)) = k\frac{\phi(B)}{\phi(A)}\lambda(A) > k\omega\lambda(B) \geq k\omega\lambda(B_0)$$

which means that T is not $k\omega$-λ-contractive.

(d) If $\omega < \delta/a$, then there is a bounded and non precompact subset B of X such that

$$\omega < \frac{\phi(B)}{\lambda(B)}\delta \leq \frac{\delta}{a}$$

and there is a ϕ-minimal set A with $\phi(A)\lambda(B)\omega < \lambda(A)\phi(B)$. We can assume that A is λ-minimal and that $A = \{x_n : n \in \mathbb{N}\}$ and moreover, since X is separable, there exists a dense and countable subset $B_0 \subset B$, that is, $B_0 = \{y_n : n \in \mathbb{N}\}$ and $\overline{B_0} = B$. Furthermore, reasoning as above, we can suppose that A is countably infinite and discrete.

If we now take $k > 0$, the mapping $T : A \to X$ defined by

$$Tx_n = k\frac{\lambda(A)}{\lambda(B)}y_n$$

is continuous, and making use of the properties of λ and ϕ, for every infinite subset C of A, we obtain

$$\lambda(T(C)) \leq \lambda(T(A)) = k\frac{\lambda(A)}{\lambda(B)}\lambda(B_0) = k\frac{\lambda(A)}{\lambda(B)}\lambda(\overline{B_0})$$

$$= k\frac{\lambda(A)}{\lambda(B)}\lambda(B) = k\lambda(A) = k\lambda(C)$$

and so T is k-λ-contractive. However

$$\phi(T(A)) = k\frac{\lambda(A)}{\lambda(B)}\phi(B_0) = k\frac{\lambda(A)}{\lambda(B)}\phi(\overline{B_0}) = k\frac{\lambda(A)}{\lambda(B)}\phi(B) > k\omega\phi(A)$$

and so T is not $k\omega$-ϕ-contractive and the proof is complete.   □

As a consequence of Theorem 2.3 we obtain the following result.

COROLLARY 2.4. *If X, λ and ϕ satisfy the conditions of the previous theorem, and if $T : D \subset X \to X$ is a continuous operator, then*

$$\frac{1}{\gamma}\lambda(T) \leq \phi(T) \leq \frac{\delta}{a}\lambda(T).$$

Moreover, if X is a separable Banach space and λ and ϕ are semi-homogeneous in X, then these constants are the best possible.

Remark 2.5. This theorem permit us to improve, in certain spaces, the trivial relationships obtained in Lemma 1.2 between the ϕ-contractiveness constants of a mapping for several MNCs. Thus we obtain:

(a) If X is a metric space, then $\gamma(\beta, \alpha)(X) = 1$ and so every k-α-contractive mapping is k-β-contractive.

(b) Less trivially, in Theorem II.4.3 we defined the MNC ν in Banach spaces with Schauder basis, and proved that $\frac{1}{L}\nu(B) \leq \chi(B) \leq \nu(B)$ for any bounded subset B, where $L = \limsup_{n \to \infty} \|R_n\|$. This result inmediately implies that if T is k-ν-contractive, then T is kL-χ-contractive. We are going to prove that if X is a reflexive Banach space with Schauder basis, then $\gamma(\chi, \nu) \leq 2 - \Delta_{X,\chi}(1^-)$. Thus, if X satisfies the uniform Opial condition, from Theorems 2.3 and V.3.11, we will be able to conclude that every k-ν-contractive mapping from a subset D of X into X is k-χ-contractive.

THEOREM 2.6. *Let X be a reflexive Banach space with Schauder basis and B a χ-minimal and ν-minimal subset of X. Then*

$$\frac{1}{2 - \Delta_{X,\chi}(1^-)}\nu(B) \leq \chi(B).$$

Proof. By translation, reflexivity, ν-minimality and multiplication we can assume, without loss of generality, that $B = \{x_n : n \in \mathbb{N}\}$, $\chi(B) = 1$ and $\{x_n\}$ is weakly null. Denote $\nu(B)$ by a and let $\{T_n\}$ be a subsequence of $\{R_n\}$ such that $\lim_{n \to \infty} \sup\{\|T_n x\| : x \in B\} = a$. For any positive number ε, $0 < \varepsilon < a/2$, we inductively construct a subsequence of $\{T_n\}$ and a subsequence of $\{x_n\}$ as follows: We choose $j_1 = 1$ and for some nonnegative number a_1, let $\{x_{1,n}\}$ be a subsequence of $\{x_n\}$ such that $|a_1 - \|T_{j_1} x_{1,n}\|| < \varepsilon$ for every $n \in \mathbb{N}$. If $j_1, j_2, \ldots, j_{k-1}$ and $\{x_{i,n}\}$ for $i = 1, 2, \ldots, k-1$ are constructed, $j_k > j_{k-1}$ is chosen such that $\|T_{j_k} x_{i,i}\| < \varepsilon$ for $i = 1, 2, \ldots, k-1$ and a subsequence $\{x_{k,n}\}$ of $\{x_{k-1,n}\}$ is chosen such that for some number a_k we have $|a_k - \|T_{j_k} x_{k,n}\|| < \varepsilon/k$. Consider the diagonal sequence $\{x_{n,n}\}$ and let k be an arbitrary positive integer. Since $\|T_{j_k} x_{n,n}\| < a/2$ if $n < k$ and $|\|T_{j_k} x_{n,n}\| - a_k| < \varepsilon/k$ if $n \geq k$, we have

$$\max\left\{\frac{a}{2}, a_k + \frac{\varepsilon}{k}\right\} \geq \liminf_{k \to \infty} \sup\{\|T_{j_k} x_{n,n}\| : n \in \mathbb{N}\} \geq \liminf_{k \to \infty} \sup\{\|T_k x_{n,n}\| : n \in \mathbb{N}\}$$

$$\geq \liminf_{k \to \infty} \sup\{\|R_k x_{n,n}\| : n \in \mathbb{N}\} = \nu(B) = a.$$

Thus $\liminf_{k \to \infty}\{a_k\} = a$. We choose k such that $|a_k - a| < \varepsilon/2$. Since

$$\chi(T_{j_k}(\{x_{n,n} : n \in \mathbb{N}\})) = \chi(\{x_{n,n} : n \in \mathbb{N}\}) = 1,$$

there exists a subset $C = \{y_n : n \in \mathbb{N}\} \subset T_{j_k}(\{x_{n,n} : n \in \mathbb{N}\})$ such that $\chi(C) = 1$ and for some $v \in X$ we have that C is contained in the ball $B(v, 1+\varepsilon)$. We denote by D the set $(1+\varepsilon)^{-1}(C-v) = \{(1+\varepsilon)^{-1}(y_n-v) : n \in \mathbb{N}\}$. Then $\chi(D) = (1+\varepsilon)^{-1}$, D is contained in $B(0,1)$ and the sequence $\{(1+\varepsilon)^{-1}(y_n-v)\}$ is weakly convergent to $-(1+\varepsilon)^{-1}v$. From Theorem V.1.11 we have $\Delta_{X,\chi}(\varepsilon') \leq 1 - (1+\varepsilon)^{-1}\|v\|$ for every $\varepsilon' < (1+\varepsilon)^{-1}$, which implies $\|v\| \leq (1+\varepsilon)(1 - \Delta_{X,\chi}(\varepsilon'))$. Thus for every $n \in \mathbb{N}$ we have

$$\|y_n\| \leq \|y_n - v\| + \|v\| \leq (1+\varepsilon)(2 - \Delta_{X,\chi}(\varepsilon')).$$

So there exists $n > k$ such that

$$a - \varepsilon < a_k - \frac{\varepsilon}{2} < \|T_{j_k}x_{n,n}\| \leq (1+\varepsilon)(2 - \Delta_{X,\chi}(\varepsilon')).$$

Since ε and $\varepsilon' < (1+\varepsilon)^{-1}$ are arbitrary, we have $a \leq 2 - \Delta_{X,\chi}(1^-)$. □

Remark 2.7. From Theorem 2.6 it follows that $\delta'(\chi,\nu)(X) \geq 1/(2 - \Delta_{X,\chi}(1^-))$. Moreover as $\delta(\chi,\nu)(X) \leq 1$, we obtain $\gamma(\chi,\nu)(X) \leq 2 - \Delta_{X,\chi}(1^-)$.

Remark 2.8. The inequality of Theorem 2.6 is not, in general, true for the measure μ defined in Theorem II.4.2. Indeed, let $|\cdot|$ be the norm in $\mathbb{R}^2$ whose unit ball is the absolutely convex hull of the vectors $(1,0)$, $(0,1)$ and $(3,3)$. Denote this Banach space by E and let $X = \ell^2(E)$, that is, $X = \{(x^n) : x^n \in E$ and $\sum_{n\geq 1}|x^n|^2 < +\infty\}$ with the norm $\|(x^n)\| = (\sum_{n\geq 1}|x^n|^2)^{1/2}$. A Schauder basis for this space is formed by the vectors $u_{2k-1} = (\delta_{nk}(1,0))$ and $u_{2k} = (\delta_{nk}(0,1))$ where $\delta_{nk} = 0$ if $n \neq k$ and $\delta_{nk} = 1$ if $n = k$. Consider the bounded set $B = \{u_{2k-1}+u_{2k} : k \geq 1\}$. It is clear that B is minimal for χ, μ and ν and that for $n > 1$ we have $\sup\{\|R_{2n-1}x\| : x \in B\} = 1/3$ and $\sup\{\|R_{2n}x\| : x \in B\} = 1$. Thus $\nu(B) = 1/3 < 1 = \mu(B)$. Since $\chi(B) = 1/3$ and it is also clear that $L = 3$, the best inequality that we can obtain for μ is $\mu(B) \leq L\chi(B)$. On the other hand, if $\{x_n\}$ is a sequence in $B(0,1)$ which converges weakly to w we can assume, taking a subsequence if necessary, that the supports of $x_n - w$ and x_n are nearly disjoint (see proof of Theorem V.1.16). Thus for every $\varepsilon > 0$ we have $1+\varepsilon \geq \limsup_{n\to\infty}\|x_n\|^2 + \varepsilon \geq \limsup_{n\to\infty}\|x_n - w\|^2 + \|w\|^2$. If $\chi(\{x_n : n \geq 1\}) \geq 1$ we have $\limsup_{n\to\infty}\|x_n - w\|^2 \geq 1$. Hence $\|w\| \leq \varepsilon$ and since ε is arbitrary, $\Delta_{X,\chi}(1^-) = 1$. Thus the inequality which we have obtained in Theorem 2.6 is the best possible in this space. Nevertheless it does not hold for μ.

Remark 2.9. It must be noted that the inequality in Theorem 2.6 is independent of the chosen Schauder basis. In contrast, the standard inequality for μ (or ν), that is, $\frac{1}{L}\mu(B) \leq \chi(B)$ depends on L and "a fortiori" on the Schauder basis. For $X = \ell^p$, since $\Delta_{X,\chi}(1^-) = 1$ we have $\chi(B) = \nu(B)$ for every χ and ν-minimal set B and for any considered Schauder basis. Obviously, if we consider the canonical basis, this result is well known because in this case $\chi = \mu = \nu$ (see Section II.4).

3. Connections between the packing rates and the normal structure coefficients

The results obtained in Chapter III about minimal sets for a measure of noncompactness permit us now to obtain some connections between the packing rates of a Banach space and the weakly convergent sequence coefficient introduced in Definition VI.3.2.

We start with an easy and useful lemma.

LEMMA 3.1. *Let X be a Banach space and $\{x_n\}$ a bounded sequence in X. Suppose that $\Phi(z) = \lim_{n\to\infty} \|x_n - z\|$ exists for all $z \in X$ and the infimum $\inf\{\Phi(z) : z \in X\}$ is attained. If $v \in X$ minimizes Φ, then $\Phi(v) = \chi(\{x_n : n \in \mathbb{N}\})$.*

Proof. Since $\Phi(v) = \lim_{n\to\infty} \|x_n - v\|$, it follows that for every $\varepsilon > 0$ there exists $n_0 \in \mathbb{N}$ such that $x_n \in B(v, \Phi(v) + \varepsilon)$ for all $n \geq n_0$, and hence $\chi(\{x_n : n \in \mathbb{N}\}) \leq \Phi(v)$.

Conversely, suppose that $\{x_n\}$ can be covered by finitely many balls with radius $r < \Phi(v)$. Then there is a ball $B(u, r)$ containing infinitely many elements of this sequence. We write the subsequence contained in this ball again as $\{x_n\}$. Then $\Phi(u) = \lim_{n\to\infty} \|x_n - u\| \leq r < \Phi(v)$ contradicting the minimality of v. $\square$

If $\{x_n\}$ is a sequence in a separable Banach space, we have proved in Lemma IX.2.5 that there is a subsequence $\{y_n\}$ such that $\lim_{n\to\infty} \|y_n - z\|$ exists for every $z \in X$. The following lemma provides a wide class of Banach spaces where $\inf\{\Phi(z) : z \in X\}$ is attained. First of all, we need remember some classic results about duality mapping (which can be found for example in [M, Chapter II]).

We shall denote $J(x)$ the *duality mapping* defined in Section V.3 for $\varphi(t) = t$. It is not difficult to prove that if X is a uniformly smooth Banach space, then $J(x)$ consists of exactly one member for all $x \in X$ and the mapping J of X into X^* is uniformly continuous on each bounded subset of X.

DEFINITION 3.2. *Let $\varphi : X \to \mathbb{R}$ a convex function. We define the subdifferential of φ at a point $z \in X$ to be the subset $\partial\varphi(z)$ of X^* given by*

$$\partial\varphi(z) = \{x^* \in X^* : \varphi(z + x) \geq \varphi(z) + x^*(x) \ \forall x \in X\}.$$

It is known that $J(z)$ is the subdifferential of the convex function $\varphi(x) = \frac{1}{2}\|x\|^2$ at every point z of X, and so we have for all x and z in X

$$x^*(x) \leq \frac{1}{2}\|x + z\|^2 - \frac{1}{2}\|z\|^2$$

for all $x^* \in J(z)$.

Finally we recall that the duality mapping is sequentially continuous at zero from ℓ^p into ℓ^q if we consider ℓ^p and ℓ^q endowed with their weak topologies and $1 < p < +\infty$, $\frac{1}{p} + \frac{1}{q} = 1$.

The following technical lemma due to Webb [W2] will be very useful in the next sections.

LEMMA 3.3. *Let X be a uniformly convex and uniformly smooth Banach space. Let $\{x_n\}$ be a sequence in X such that $\Phi(z) = \lim_{n\to\infty} \|x_n - z\|$ exists for all $z \in X$. Then, there is a unique point $v \in X$ such that $\{J(x_n - v)\}$ is weakly convergent to zero in X^*. In fact, v is the point of X where Φ attains its unique absolute minimum.*

Proof. From Lemma VI.3.7 we know that Φ is lower semicontinuous for the weak topology. Hence, Φ attains an absolute minimum over every weak compact set in X. Since $\Phi(z) \to +\infty$ when $\|z\| \to +\infty$ we can conclude that Φ attains an absolute minimum over X at a point v.

Let us see that this minimum is unique. Indeed, suppose there are two points v_1 and v_2 where Φ attains its absolute minimum and let ε be a positive real number.

Then there is $n_0 \in \mathbb{N}$ such that for all $n \geq n_0$ we have

$$\|x_n - v_1\| \leq \Phi(v_1) + \varepsilon$$

and

$$\|x_n - v_2\| \leq \Phi(v_2) + \varepsilon.$$

Let $R = \Phi(v_1) + \varepsilon = \Phi(v_2) + \varepsilon > 0$. Then, as $\|x_n - v_1\| \leq R$, $\|x_n - v_2\| \leq R$ and the space X is uniformly convex, we can conclude that

$$\left\| x_n - \frac{v_1 + v_2}{2} \right\| \leq R\left(1 - \delta_X\left(\frac{\|v_1 - v_2\|}{R}\right)\right)$$

for all $n \geq n_0$. Hence

$$\Phi\left(\frac{v_1 + v_2}{2}\right) = \lim_{n\to\infty} \left\| x_n - \frac{v_1 + v_2}{2} \right\| \leq R\left(1 - \delta_X\left(\frac{\|v_1 - v_2\|}{R}\right)\right)$$

and now letting $\varepsilon \to 0$ we obtain

$$\Phi\left(\frac{v_1 + v_2}{2}\right) \leq \Phi(v_1)\left(1 - \delta_X\left(\frac{\|v_1 - v_2\|}{R}\right)\right) < \Phi(v_1)$$

contradicting the minimality of v_1.

Let us see now that $\{J(x_n - v)\}$ is weakly convergent to zero. Indeed, as $J(z)$ is the subdifferential of the convex function $\frac{1}{2}\|x\|^2$ in every point z of X, we have

$$J(x_n - z)(z - v) \leq \frac{1}{2}\|x_n - v\|^2 - \frac{1}{2}\|x_n - z\|^2. \tag{1}$$

We fix $t > 0$ and $u \in X$ and let $z_t = v + tu$. We have $\limsup_{n\to\infty} J(x_n - z_t)$ $(tu) \leq (1/2)\limsup_{n\to\infty} \left(\|x_n - v\|^2 - \|x_n - v - tu\|^2\right) \leq 0$, because $\Phi(v)$ is the infimum. Cancel $t > 0$ and then let $t \to 0$. As J is uniformly continuous on bounded sets, we obtain $\limsup_{n\to\infty} J(x_n - v)(u) \leq 0$. Replacing u by $-u$ we again obtain $\limsup_{n\to\infty} J(x_n - v)(-u) \leq 0$ and so $\lim_{n\to\infty} J(x_n - v)(u) = 0$ for every $u \in X$. It follows that $\{J(x_n - v)\}$ converges weakly to zero.

Finally we show that if $\{J(x_n - w)\}$ converges weakly to zero, then $v = w$. Indeed, we take $z = w$ in (1) and obtain

$$J(x_n - w)(w - v) \leq \frac{1}{2}\|x_n - v\|^2 - \frac{1}{2}\|x_n - w\|^2$$

and now letting $n \to +\infty$ we obtain

$$0 \leq \frac{1}{2}\lim_{n\to\infty}\|x_n - v\|^2 - \frac{1}{2}\lim_{n\to\infty}\|x_n - w\|^2$$

and hence

$$\lim_{n\to\infty}\|x_n - w\|^2 \leq \lim_{n\to\infty}\|x_n - v\|^2.$$

It follows that $v = w$ because Φ attains at v its unique absolute minimum. ∎

PROPOSITION 3.4. *Let X be an infinite dimensional, separable and reflexive Banach space. Then the following inequality holds:*

$$WCS(X) \leq \frac{1}{\delta(\chi, \alpha)(X)}.$$

Proof. Let B be a bounded, α-minimal and nonprecompact subset of X. From Lemma III.2.7, there exists a χ-minimal subset B_1 of B such that $\chi(B_1) = \chi(B)$. As B_1 is χ and α-minimal, we can suppose that $B_1 = \{x_n : n \in \mathbb{N}\}$ with $x_n \neq x_m$ for $n \neq m$ and bearing in mind Theorem III.1.5 we can also suppose that $\lim_{n,m\,;n\neq m}\|x_n - x_m\| = \alpha(B_1)$. Since X is separable it follows from Lemma IX.2.5, taking a subsequence if necessary, that $\Phi(z) = \lim_{n\to\infty}\|x_n - z\|$ exists for all $z \in X$. Furthermore, as X is reflexive and Φ is a lower semicontinuous function, the infimum $\inf\{\Phi(z) : z \in X\}$ is attained (see proof of Lemma 3.3). If $v \in X$ minimizes Φ, from Lemma 3.1 we know $\Phi(v) = \chi(\{x_n\})$ and so $\chi(\{x_n\}) \leq \inf\{\Phi(z) : z \in \overline{co}(\{x_n\})\}$. Thus

$$\frac{\chi(B)}{\alpha(B)} = \frac{\chi(B_1)}{\alpha(B_1)} \leq \frac{\inf\{\Phi(z) : z \in \overline{co}(\{x_n\})\}}{\operatorname{diam}_a(x_n)} \leq \frac{1}{WCS(X)}$$

and so

$$\delta(\chi, \alpha)(X) \leq \frac{1}{WCS(X)}.$$

∎

PROPOSITION 3.5. *Let X be a separable, reflexive and infinite dimensional Banach space with Opial's condition. Then $WCS(X) = 1/\delta(\chi, \alpha)(X)$.*

Proof. Let $\{x_n\}$ be a normalized weakly null sequence in X such that $\lim_{n,m\,;n\neq m}$ $\|x_n - x_m\| = l$. Consider the set $A = \{x_n : n \in \mathbb{N}\}$. It is clear that A is α-minimal, $\alpha(A) = l$ and $\chi(A) \leq 1$. The Opial's condition implies $\chi(A) = 1$. Indeed, if $\chi(A) < r < 1$ there exists $z \in X$ such that $B(z,r)$ contains infinitely many points of $\{x_n\}$. Thus $\liminf_{n\to\infty} \|x_n - z\| \leq r < 1 = \lim_{n\to\infty} \|x_n\|$, contradicting Opial's condition. Hence we have

$$\delta(\chi,\alpha)(X) \geq \frac{\chi(A)}{\alpha(A)} = \frac{1}{\alpha(A)} = \frac{1}{\lim_{n,m\,;n\neq m} \|x_n - x_m\|}$$

and it follows from Lemma VI.3.8 that $\delta(\chi,\alpha)(X) \geq 1/WCS(X)$.   □

4. Packing rates in ℓ^p-spaces

In this section we shall compute the packing rates in ℓ^p-spaces, $1 \leq p < +\infty$. Previously, we need two technical lemmas.

LEMMA 4.1. *Let* $\{x_n\} = \{(x_n^k)\}$ *a sequence in* ℓ^p, $1 \leq p < +\infty$, *such that* $\lim_{n\to\infty} x_n^k = 0$ *for each* $k \in \mathbb{N}$. *Assume that there are constants* a_1, a_2, α_1, α_2 *such that* $a_1 \leq \|x_n\| \leq a_2$ *and* $\alpha_1 \leq \|x_n - x_1\| \leq \alpha_2$ *for every* $n \in \mathbb{N}$. *Then* $2^{\frac{1}{p}} a_1 \leq \alpha_2$ *and* $\alpha_1 \leq 2^{\frac{1}{p}} a_2$.

Proof. By a standard argument (see, for instance, the proof of Theorem V.1.16) we can construct sequences $\{u_n\}$, $\{v_n\}$, $\{y_n\}$ and $\{z_n\}$ such that for every $n \in \mathbb{N}$ we have $x_1 = u_n + v_n$, $x_n = y_n + z_n$, $\lim_{n\to\infty} \|v_n\| = \lim_{n\to\infty} \|y_n\| = 0$ and $\mathrm{supp}(u_n) \cap \mathrm{supp}(z_n) = \emptyset$ for every $n > 1$.

Let ε be an arbitrary positive number. For $n \in \mathbb{N}$ large enough we have $\alpha_1 - \varepsilon \leq \|z_n - u_n\| \leq \alpha_2 + \varepsilon$, $a_1 - \varepsilon \leq \|z_n\| \leq a_2 + \varepsilon$ and $a_1 - \varepsilon \leq \|u_n\| \leq a_2 + \varepsilon$. Therefore

$$(\alpha_1 - \varepsilon)^p \leq \|z_n - u_n\|^p = \|z_n\|^p + \|u_n\|^p \leq 2(a_2 + \varepsilon)^p$$

and

$$(\alpha_2 + \varepsilon)^p \geq \|z_n - u_n\|^p = \|z_n\|^p + \|u_n\|^p \geq 2(a_1 - \varepsilon)^p.$$

Since $\varepsilon > 0$ was chosen arbitrarily, we obtain the required inequalities.   □

LEMMA 4.2. *Let* $\{x_n\} = \{(x_n^k)\}$ *a sequence in* ℓ^p, $1 \leq p < +\infty$, *such that* $\lim_{n\to\infty} x_n^k = v^k$ *for each* $k \in \mathbb{N}$ *and there exists* $\Phi(z) = \lim_{n\to\infty} \|x_n - z\|$ *for every* $z \in \ell^p$. *Then* $v = (v^k)$ *belongs to* ℓ^p *and for every* $z \in \ell^p$ *one has* $\Phi(z) > \Phi(v)$. *Furthermore,* $\chi(\{x_n : n \in \mathbb{N}\}) = \Phi(v)$.

Proof. Case $p > 1$. Taking a subsequence with the same χ-measure if necessary, we obtain that $\{x_n\}$ converges weakly to v, and so, v belongs to ℓ^p. Since the duality mapping $J : \ell^p \to \ell^q$ is sequentially continuous at zero when ℓ^p and ℓ^q are endowed with the weak topologies, we have that $J(x_n - v)$ converges weakly to zero. By Lemma 3.3 Φ attains its unique absolute minimum at v.

Case $p = 1$. It is easy to check that v belongs to ℓ^1. Assume that there exists $w \in \ell^1$ such that $\Phi(w) < \Phi(v)$ and write $\varepsilon = (\Phi(v) - \Phi(w))/5$. Now choose $m \in \mathbb{N}$ such that

$$\sum_{k=m+1}^{\infty} |w^k| < \varepsilon, \qquad \sum_{k=m+1}^{\infty} |v^k| < \varepsilon$$

and $n \in \mathbb{N}$ such that

$$\left| \|x_n - w\| - \Phi(w) \right| < \varepsilon, \quad \left| \|x_n - v\| - \Phi(v) \right| < \varepsilon, \quad |x_n^k - v^k| < \frac{\varepsilon}{m}$$

for $k = 1, 2, \ldots, m$. Then

$$\Phi(v) < \|x_n - v\| + \varepsilon = \sum_{k=1}^{m} |x_n^k - v^k| + \sum_{k=m+1}^{\infty} |x_n^k - v^k| + \varepsilon$$

$$\leq 2\varepsilon + \sum_{k=m+1}^{\infty} |x_n^k - w^k| + \sum_{k=m+1}^{\infty} |v^k| + \sum_{k=m+1}^{\infty} |w^k|$$

$$\leq 4\varepsilon + \|x_n - w\| < 5\varepsilon + \Phi(w) = \Phi(v).$$

This contradiction proves $\Phi(v) \leq \Phi(z)$ for every $z \in \ell^1$. Furthermore, it is easy to check that this inequality is strict. Indeed, assume as a contradiction, that $\Phi(v) = \Phi(w)$ for some $w \in \ell^1$ and let $w^k = v^k$ for $k = 1, 2, \ldots, h-1$ and $w^h \neq v^h$. Write $\varepsilon = \frac{|w^h - v^h|}{3}$ and n_0 is chosen such that $|x_n^h - v^h| < \varepsilon$ for every $n \geq n_0$. Denote $u = \sum_{k \neq h} w^k e_k + v^h e_h$. Then for every $n \geq n_0$ we have

$$\|x_n - w\| = \sum_{k \neq h} |x_n^k - w^k| + |x_n^h - w^h|$$

$$\geq \sum_{k \neq h} |x_n^k - w^k| + |w^h - v^h| + |x_n^h - v^h|$$

$$\geq \sum_{k \neq h} |x_n^k - w^k| + 2\varepsilon$$

$$= \|x_n - u\| + 2\varepsilon - |x_n^h - v^h| \geq \|x_n - u\| + \varepsilon.$$

Taking limits as $n \to +\infty$ we obtain $\Phi(v) = \Phi(w) \geq \Phi(u) + \varepsilon > \Phi(u)$, contradicting the minimality of v.

Finally, the equality $\chi(\{x_n : n \in \mathbb{N}\}) = \Phi(v)$ follows from Lemma 3.1 for $p \geq 1$. $\qquad \square$

Remark.4.3. A dual space X is said to satisfy the weak* Opial's condition [Ka2] if

$$\lim_{n\to\infty} \|x_n - z\| > \lim_{n\to\infty} \|x_n - v\|$$

for every sequence $\{x_n\}$ in X weakly* convergent to v. Obviously weak* Opial's condition and Opial's condition are identical if X is reflexive. In the proof of Lemma 4.2 we have shown that every ℓ^p, $1 \le p < +\infty$, has the weak* Opial's condition. This result was first proved in [Li1].

THEOREM 4.4. *Let $X = \ell^p$ with $1 \le p < +\infty$. Let A be an infinite subset of X satisfying $\alpha_1 \le \|x - y\| \le \alpha_2$ for every $x,y \in A$, $x \ne y$. Then $2^{\frac{-1}{p}}\alpha_1 \le \chi(A) \le 2^{\frac{-1}{p}}\alpha_2$.*

Proof. Let B be a χ-minimal subset of A such that $\chi(B) = \chi(A)$ (see Theorem III.2.7) and assume that $B = \{x_n : n \in \mathbb{N}\}$. By using a diagonal method we can find a subsequence $\{y_n\}$ of $\{x_n\}$ such that $\lim_{n\to\infty} y_n^k = v^k$ for every $k \in \mathbb{N}$. Since X is separable, we can also assume that $\lim_{n\to\infty} \|y_n - z\| = \Phi(z)$ exists for every $z \in \ell^p$ (see Lemma IX.2.5). From Lemma 4.2 we know that $\Phi(v) = \chi(\{y_n : n \in \mathbb{N}\}) = \chi(A)$.

Let ε be an arbitrary positive number. We can assume that

$$\chi(A) - \varepsilon \le \|y_n - v\| \le \chi(A) + \varepsilon$$

for all $n \in \mathbb{N}$. Since $\alpha_1 \le \|y_n - y_m\| \le \alpha_2$, by applying Lemma 4.1 to the sequence $\{y_n - v\}$ we obtain

$$2^{\frac{1}{p}}(\chi(A) - \varepsilon) \le \alpha_2; \quad \alpha_1 \le 2^{\frac{1}{p}}(\chi(A) + \varepsilon)$$

and letting $\varepsilon \to 0$ we conclude $\alpha_1 \le 2^{\frac{1}{p}}\chi(A) \le \alpha_2$. $\qquad\square$

COROLLARY 4.5. *Let $1 \le p < +\infty$. Then:*

(a) $2^{\frac{-1}{p}}\alpha(A) = \chi(A)$ for every α-minimal subset A of ℓ^p.

(b) $\delta(\chi, \alpha)(\ell^p) = \delta'(\chi, \alpha)(\ell^p) = 2^{\frac{-1}{p}}$ and so $\gamma(\chi, \alpha)(\ell^p) = 1$.

Proof. (a) Let B be an α-minimal and χ-minimal subset of A such that $\chi(A) = \chi(B)$. Given $\varepsilon > 0$, we can find an infinite subset C of B such that

$$\alpha(A) - \varepsilon = \alpha(B) - \varepsilon \le \|x - y\| \le \alpha(B) + \varepsilon \le \alpha(A) + \varepsilon$$

for every $x, y \in C$, $x \ne y$ (see Lemma III.1.3). Bearing in mind Theorem 4.4 and $\chi(C) = \chi(B) = \chi(A)$, we obtain

$$2^{\frac{-1}{p}}(\alpha(A) - \varepsilon) \le \chi(A) \le 2^{\frac{-1}{p}}(\alpha(A) + \varepsilon).$$

Letting $\varepsilon \to 0$ we reach that $2^{\frac{-1}{p}}\alpha(A) = \chi(A)$.

(b) It follows immediately from (a). $\qquad\square$

Remark 4.6. In Corollary 4.5 we have proved that if A is an α-minimal subset of ℓ^1, then $\alpha(A) = 2\chi(A)$. Actually, it is easy to check that $\alpha(A) = 2\chi(A)$ for every bounded subset A of ℓ^1. This is also the situation for ℓ^∞ (see Example II.2).

COROLLARY 4.7. *Let* $1 \leq p < +\infty$ *and* A *a bounded subset of* ℓ^p. *Then* $\beta(A) = 2^{\frac{1}{p}}\chi(A)$.

Proof. Bearing in mind Lemma III.2.8, Theorem III.1.2, Corollary 4.5 (a) and Theorem III.2.7, we can write the following equalities:

$$\beta(A) = \sup\{\alpha(B) : B \subset A, \ B \ \alpha\text{-minimal}\}$$
$$= \sup\{2^{\frac{1}{p}}\chi(B) : B \subset A, \ B \ \alpha\text{-minimal}\}$$
$$= 2^{\frac{1}{p}}\chi(A).$$

$\square$

Using to the above results, we deduce the following values for the packing rates associated with the MNCs α, β and χ in ℓ^p-spaces ($1 \leq p < +\infty$):

$$\delta(\chi,\beta) = \delta'(\chi,\beta) = 2^{\frac{-1}{p}}; \ \ \delta(\beta,\chi) = \delta'(\beta,\chi) = 2^{\frac{1}{p}}$$
$$\delta(\chi,\alpha) = \delta'(\chi,\alpha) = 2^{\frac{-1}{p}}; \ \ \delta'(\alpha,\chi) = 2^{\frac{1}{p}}$$
$$\delta(\beta,\alpha) = \delta'(\beta,\alpha) = 1; \ \ \delta'(\alpha,\beta) = 1$$

and so

$$\gamma(\chi,\beta) = \gamma(\chi,\alpha) = \gamma(\beta,\chi) = \gamma(\beta,\alpha) = 1.$$

Moreover, for the spaces ℓ^p ($1 \leq p < +\infty$) these results permit the best relationships to be reached between ϕ-contractive operators associated to these MNCs.

5. Packing rates in L^p-spaces

In this section we are going to calculate the packing rates for χ and α in infinite dimensional L^p-spaces with a not purely atomic measure. As a consequence, the remaining coefficients for the three main MNCs will be derived.

LEMMA 5.1. *Let* α_1, α_2 *be real numbers such that* $\alpha_1 \leq \alpha_2$. *Let* $\{x_n\}$ *be a sequence in* $L^p(\Omega)$, *where* (Ω, Σ, μ) *is a σ-finite measure space and* $1 < p < +\infty$. *Assume that* $\alpha_1 \leq \|x_n - x_m\| \leq \alpha_2$ *for each* $n, m \in \mathbb{N}$, $n \neq m$, *and let* $a = \chi(\{x_n : n \in \mathbb{N}\})$. *If* $\frac{1}{p} + \frac{1}{q} = 1$, *then:*

(a) $\alpha_1 \leq 2^{\frac{1}{p}}a$ *and* $\alpha_2 \geq 2^{\frac{1}{q}}a$ *if* $1 < p \leq 2$.

(b) $\alpha_1 \leq 2^{\frac{1}{q}}a$ *and* $\alpha_2 \geq 2^{\frac{1}{p}}a$ *if* $2 \leq p < +\infty$.

Proof. We can assume without loss of generality that $\{x_n : n \in \mathbb{N}\}$ is an α-minimal and χ-minimal set. Moreover, taking a subsequence if necessary, we can suppose that $\{x_n\}$ is weakly convergent and, by translation, that the weak limit of $\{x_n\}$ is zero. Since $L^p(\Omega)$ is separable, we can also assume that $\lim_{n\to\infty} \|x_n - z\| = \Phi(z)$ exists for every $z \in L^p(\Omega)$ (see Lemma IX.2.5). Furthermore, Φ attains its unique minimum at a point $v \in L^p(\Omega)$ and, in fact, $\Phi(v) = a$ (see Lemmas 3.1 and 3.3).

Let ε be a positive real number. Since $\Phi(v) = a$ we can also assume that $a - \varepsilon \leq \|x_n - v\| \leq a + \varepsilon$ for every $n \in \mathbb{N}$. Let us apply the inequalities of Lemma II.3.8 to the vectors $x_1 - v, x_2 - v, \ldots, x_n - v \in L^p(\Omega)$.

For $1 < p \leq 2$ taking $t_j = \frac{1}{n}, j = 1, 2, \ldots, n$ and $2\alpha = p$, from Lemma II.3.8 (a) we obtain

$$\frac{n-1}{n}\alpha_1^p \leq 2\left(\frac{n-1}{n}\right)^{2-p}(a+\varepsilon)^p$$

for every $n \in \mathbb{N}$. Since n can be chosen arbitrarily large we deduce

$$\alpha_1^p \leq 2(a+\varepsilon)^p$$

and letting $\varepsilon \to 0$ we conclude $\alpha_1 \leq 2^{\frac{1}{p}}a$. The same argument using Lemma II.3.8 (b) for $2 \leq p < +\infty$ with $2\alpha = q$ proves $\alpha_1 \leq 2^{\frac{1}{q}}a$.

Now let $r = \lim_{n\to\infty} \|x_n\|$ and $\varepsilon > 0$. Since $\Phi(v) \leq \Phi(0)$ it follows that $r \geq a$ and therefore we can suppose that $\|x_n\| \geq a - \varepsilon$ for every $n \in \mathbb{N}$. To obtain the inequalities concerning α_2 we use the weak convergence of $\{x_n\}$ to zero to imply that there exists $m \in \mathbb{N}$ and m positive numbers $\lambda_1, \lambda_2, \ldots, \lambda_m$ with $\sum_{i=1}^{m} \lambda_i = 1$ such that $\|\sum_{i=1}^{m} \lambda_i x_i\| < \varepsilon$. Applying Lemma II.3.8 (c) if $1 < p \leq 2$ with $\beta = q$ and $\gamma = 1$ we obtain

$$\alpha_2^q \geq 2(a - 2\varepsilon)^q.$$

Since $\varepsilon > 0$ is arbitrary we have $\alpha_2 \geq 2^{\frac{1}{q}}a$ and we analogously obtain $\alpha_2 \geq 2^{\frac{1}{p}}a$ in the case $2 \leq p < +\infty$ using Lemma II.3.8 (d). Hence the proof is complete. $\square$

THEOREM 5.2. *Let A be a bounded, α-minimal and nonprecompact subset of $L^p(\Omega)$, where $1 \leq p < +\infty$ and (Ω, Σ, μ) is a σ-finite measure space. Then*

$$\min\{2^{\frac{1}{p}-1}, 2^{\frac{-1}{p}}\} \leq \frac{\chi(A)}{\alpha(A)} \leq \max\{2^{\frac{1}{p}-1}, 2^{\frac{-1}{p}}\}.$$

Moreover, these bounds are the best possible if μ is not purely atomic.

Proof. For $p = 1$ the inequalities are obvious. Assume $1 < p < +\infty$ and let A be a bounded, α-minimal and nonprecompact subset of $L^p(\Omega)$. Then there exists an α-minimal and χ-minimal subset B of A such that $\chi(B) = \chi(A)$. We can also suppose that $B = \{x_n : n \in \mathbb{N}\}$.

Moreover, for each $\varepsilon > 0$ there exists an infinite subset of B, again denoted by B, such that

$$\alpha(A) - \varepsilon \leq \|x_n - x_m\| \leq \alpha(A) + \varepsilon \text{ for each } x_n, x_m \in B, \quad n \neq m$$

(see Lemma III.1.3). From Lemma 5.1 it follows that:

If $1 < p \leq 2$ then $\alpha(A) - \varepsilon \leq 2^{\frac{1}{p}}\chi(A)$ and since $\varepsilon > 0$ is arbitrary, we obtain $\alpha(A) \leq 2^{\frac{1}{p}}\chi(A)$, that is, $2^{\frac{-1}{p}} \leq \frac{\chi(A)}{\alpha(A)}$.

On the other hand $\alpha(A) + \varepsilon \geq 2^{1-\frac{1}{p}}\chi(A)$. Hence $2^{\frac{1}{p}-1} \geq \frac{\chi(A)}{\alpha(A)}$. Thus if $1 < p \leq 2$ we have $2^{\frac{-1}{p}} \leq \frac{\chi(A)}{\alpha(A)} \leq 2^{\frac{1}{p}-1}$.

Analogously, if $2 \leq p < +\infty$ we obtain $2^{\frac{1}{p}-1} \leq \frac{\chi(A)}{\alpha(A)} \leq 2^{\frac{-1}{p}}$.

Finally, we show that the bounds are attained if μ is not purely atomic. Indeed, the argument in the proof of Lemma II.3.9 proves that we can construct a sequence $\{A_n\}$ of measurable subsets of Ω with $\mu(A_n) > 0$, $A_n \cap A_m = \emptyset$ for all $n \neq m$ and $\mu(A_n) \to 0$ when $n \to \infty$.

For $1 \leq p < +\infty$ we define

$$x_{n,p} = \mu(A_n)^{\frac{-1}{p}} \chi_{A_n}.$$

It is easy to check that $\|x_{n,p} - x_{m,p}\| = 2^{\frac{1}{p}}$ for all $n, m \in \mathbb{N}$, $n \neq m$, and that $\|x_{n,p}\| = 1$ for all $n \in \mathbb{N}$, where norms are considered in the corresponding space $L^p(\Omega)$. Therefore, the set $A = \{x_{n,p} : n \in \mathbb{N}\}$ is a bounded, α-minimal subset of $L^p(\Omega)$ with $\alpha(A) = 2^{\frac{1}{p}}$. Moreover, obviously $\chi(A) \leq 1$. We will now prove that, in fact, $\chi(A) = 1$.

Let $x \in L^p(\Omega)$. Using the Hölder inequality we obtain

$$\left| \int_\Omega x_{n,q}(t)x(t)dt \right| = \left| \int_{A_n} x_{n,q}(t)x(t)dt \right|$$

$$\leq \left| \int_{A_n} |x_{n,q}(t)|^q dt \right|^{\frac{1}{q}} \left| \int_{A_n} |x(t)|^p dt \right|^{\frac{1}{p}}$$

$$= \left| \int_{A_n} |x(t)|^p dt \right|^{\frac{1}{p}}$$

and the last term converges to zero because $|x|^p \in L^1(\Omega)$ and $\mu(A_n) \to 0$. Thus, for $1 \leq p < +\infty$ and $x \in L^p(\Omega)$ we have

$$
\begin{aligned}
1 &= \left| \int_\Omega x_{n,p}(t)\mu(A_n)^{\frac{-1}{q}} dt \right| = \left| \int_\Omega x_{n,p}(t)x_{n,q}(t)dt \right| \\
&\leq \left| \int_\Omega (x_{n,p}(t) - x(t))x_{n,q}(t)dt \right| + \left| \int_\Omega x(t)x_{n,q}(t)dt \right| \\
&\leq \|x_{n,p} - x\|_p \|x_{n,q}\|_q + \left| \int_\Omega x(t)x_{n,q}(t)dt \right| \\
&= \|x_{n,p} - x\| + \left| \int_\Omega x(t)x_{n,q}(t)dt \right|.
\end{aligned}
$$

Taking the limit inferior as $n \to \infty$ and bearing in mind that $\left| \int_\Omega x(t)x_{n,q}(t)dt \right| \to 0$, we obtain $1 \leq \liminf_{n\to\infty} \|x_{n,p} - x\|$ for all $x \in L^p(\Omega)$. Hence $\chi(A) = 1$. Therefore we have found a bounded, α-minimal and nonprecompact subset A of $L^p(\Omega)$ such that $\frac{\chi(A)}{\alpha(A)} = 2^{\frac{-1}{p}}$ with $1 \leq p < +\infty$.

On the other hand, since μ is not purely atomic, we can construct the set $B = \{y_{n,p} : n \in \mathbb{N}\}$ of the Rademacher functions (see proof of Theorem II.3.12) given by

$$
y_{n,p} = \mu(E)^{-\frac{1}{p}} \sum_{k=1}^{2^n} (-1)^{k+1} \chi_{E_{n,k}}
$$

where the sets $\{E_{n,j}\}_{j=1}^{2^n}$ are pairwise disjoint and verify that $\bigcup_{j=1}^{2^n} E_{n,j} = E$ and

$$
E_{n-1,k} = E_{n,2k-1} \cup E_{n,2k} \quad \text{and} \quad \mu(E_{n,2k-1}) = \mu(E_{n,2k}) = \frac{1}{2}\mu(E_{n-1,k})
$$

for $k = 1, 2, \ldots, 2^{(n-1)}$.

The elements of this sequence satisfy $\|y_{n,p}\| = 1$ for every $n \in \mathbb{N}$ and $\|y_{n,p} - y_{m,p}\| = 2^{1-\frac{1}{p}}$ for all $n \neq m$, and so B is a bounded, α-minimal and nonprecompact subset of $L^p(\Omega)$ with $\alpha(B) = 2^{1-\frac{1}{p}}$. Furthermore $\chi(B) \leq 1$ and $\lim_{n\to\infty} \int_\Omega y_{n,p}(t)x(t)dt = 0$ for any function $x \in L^1(\Omega)$. Thus the same argument as above proves that $\chi(B) = 1$. Therefore we have found a bounded, α-minimal and nonprecompact subset B of $L^p(\Omega)$ such that $\frac{\chi(B)}{\alpha(B)} = 2^{\frac{1}{p}-1}$ with $1 \leq p < +\infty$. Hence the proof is complete. $\square$

COROLLARY 5.3. *Let (Ω, Σ, μ) be a σ-finite measure space and $1 \leq p < +\infty$. Then*

$$
\delta'(\chi,\alpha)(L^p(\Omega)) \geq \min\{2^{\frac{1}{p}-1}, 2^{\frac{-1}{p}}\}; \quad \delta(\chi,\alpha)(L^p(\Omega)) \leq \max\{2^{\frac{1}{p}-1}, 2^{\frac{-1}{p}}\}
$$

and so

$$
\gamma(\chi,\alpha)(L^p(\Omega)) \leq 2^{\frac{|p-2|}{p}}.
$$

Moreover, the equalities hold if μ is not purely atomic.

Hence, the following values are obtained for the packing rates associated with the MNCs α, β and χ in $L^p(\Omega)$, $(1 \leq p < +\infty$, where (Ω, Σ, μ) is a σ-finite measure space with a not purely atomic measure μ):

$$\delta(\chi, \beta) = \max\{2^{\frac{1}{p}-1}, 2^{\frac{-1}{p}}\} = \delta(\chi, \alpha)$$

$$\delta'(\chi, \beta) = \min\{2^{\frac{1}{p}-1}, 2^{\frac{-1}{p}}\} = \delta'(\chi, \alpha)$$

$$\delta(\beta, \chi) = \max\{2^{1-\frac{1}{p}}, 2^{\frac{1}{p}}\}$$

$$\delta'(\beta, \chi) = \min\{2^{1-\frac{1}{p}}, 2^{\frac{1}{p}}\} = \delta'(\alpha, \chi)$$

$$\delta(\beta, \alpha) = \delta'(\beta, \alpha) = \delta'(\alpha, \beta) = 1$$

and so

$$\gamma(\chi, \beta) = \gamma(\chi, \alpha) = \gamma(\beta, \chi) = 2^{\frac{|p-2|}{p}}$$

$$\gamma(\beta, \alpha) = 1.$$

Furthermore, these results permit the best relationships to be obtained between ϕ-contractive operators associated with these MNCs in the above class of L^p-spaces.

The case $p = \infty$ is much easier. We have the following proposition.

PROPOSITION 5.4. *Let (Ω, Σ, μ) be a measure space. Then every bounded subset A of $L^\infty(\Omega)$ has $\alpha(A) = 2\chi(A)$.*

Proof. The inequality $\alpha(A) \leq 2\chi(A)$ holds in every metric space. Conversely, let ε be an arbitrary positive number and $A_1, A_2, \ldots, A_r$ be sets in $L^\infty(\Omega)$ such that $A \subseteq \bigcup_{i=1}^{r} A_i$ and $\operatorname{diam}(A_i) \leq \alpha(A) + \varepsilon$. Since $L^\infty(\Omega)$ is a complete Banach lattice, there exist $f_1^i = \sup\{f : f \in A_i\}$ and $f_2^i = \inf\{f : f \in A_i\}$ for all $i = 1, 2, \ldots, r$. Let $h_i = (f_1^i + f_2^i)/2$ for all $i = 1, 2, \ldots, r$. It is easy to check that $A_i \subseteq B_i = B\left(h_i, \frac{\alpha(A)+\varepsilon}{2}\right)$ for all $i = 1, 2, \ldots, r$. Thus $\chi(A) \leq (\alpha(A) + \varepsilon)/2$ and letting $\varepsilon \to 0$ we obtain that $2\chi(A) \leq \alpha(A)$ and the proof is complete. □

So, the packing rates in this case are the following:

$$\delta(\chi, \beta) = \delta'(\chi, \beta) = \frac{1}{2}; \quad \delta(\beta, \chi) = \delta'(\beta, \chi) = 2$$

$$\delta(\chi, \alpha) = \delta'(\chi, \alpha) = \frac{1}{2}; \quad \delta(\alpha, \chi) = \delta'(\alpha, \chi) = 2$$

$$\delta(\beta, \alpha) = \delta'(\beta, \alpha) = 1; \quad \delta'(\alpha, \beta) = 1$$

and so

$$\gamma(\chi, \beta) = \gamma(\chi, \alpha) = \gamma(\beta, \chi) = \gamma(\beta, \alpha) = \gamma(\alpha, \chi) = 1.$$

6. Packing rates in direct sum spaces

In this section we compute the coefficients $\delta(\chi,\alpha)(X)$, $\delta'(\chi,\alpha)(X)$ and $\gamma(\chi,\alpha)(X)$ (which will be denoted in this section as $\delta(X)$, $\delta'(X)$ and $\gamma(X)$ respectively) when X is a direct sum of separable Banach spaces. For a finite direct sum we require the substitution space to have a monotone norm. In the infinite case we only show results if the substitution space has a p-norm with $1 \leq p < +\infty$. This type of problem has been studied for other coefficients in [Ca], [Do5], [KL1], [KL2], [L1] and [L2].

We recall that a norm on $\mathbb{R}^n$ is said to be monotone if $\|(a_1,\ldots,a_k)\| \leq \|(b_1,\ldots,b_k)\|$ when $0 \leq a_i \leq b_i$ for every $i = 1,\ldots,k$. This condition is satisfied if, for instance, the norm is symmetric, that is, $\|(a_1,\ldots,a_k)\| = \|(\varepsilon_1 a_1,\ldots,\varepsilon_k a_k)\|$ for any $\varepsilon_i = \pm 1$. This is the case for the p-norms or Orlicz norms.

It is easy to prove the following lemma.

LEMMA 6.1. *Let $X_1,\ldots,X_k$ be separable Banach spaces and $|\cdot|$ a monotone norm on $\mathbb{R}^k$. Then, the product space $X = X_1 \times \cdots \times X_k$ with the norm*

$$\|(x_1,\ldots,x_k)\| = |(\|x_1\|,\ldots,\|x_k\|)|$$

is a separable Banach space which will be denoted by $X_1 \oplus \cdots \oplus X_k$.

We shall need the following fact to prove the main result in the finite case.

LEMMA 6.2. *Let A be an α-minimal set in $X = X_1 \oplus \cdots \oplus X_k$ such that its projection A_i into X_i is either an α-minimal and χ-minimal set or a finite set, for each $i = 1,\ldots,k$. Then*

$$\alpha(A) = |(\alpha(A_1),\ldots,\alpha(A_k))| \quad and \quad \chi(A) = |(\chi(A_1),\ldots,\chi(A_k))|.$$

Proof. As usual, we denote $x = (x^i) = (x^1,\ldots,x^k)$ where $x^i \in X_i$. Since $X_1 \oplus \cdots \oplus X_k$ is separable, we can suppose $A = \{x_n : n \in \mathbb{N}\} = \{(x_n^i) : n \in \mathbb{N}\}$ and, taking subsequences if necessary, we can also assume that the projections $A_i = \{x_n^i : n \in \mathbb{N}\}$ are α-minimal sets or singletons. From Theorem III.1.5, we assume that $\lim_{n,m\ n\neq m} \|(x_n^i) - (x_m^i)\| = \alpha(A)$ and $\lim_{n,m\ n\neq m} \|x_n^i - x_m^i\| = \alpha(A_i)$ for every $i = 1,\ldots,k$. So, given $\varepsilon > 0$ such that $\varepsilon < \min_{1\leq i \leq k}\{\alpha(A_i) : \alpha(A_i) > 0\}$, there exists $n_0 \in \mathbb{N}$ such that if $n > n_0, m > n_0, n \neq m$ we have $\alpha(A) - \varepsilon \leq \|(x_n^i) - (x_m^i)\| = |(\|x_n^i - x_m^i\|)| \leq |(\alpha(A_i) + \varepsilon)|$, since the norm is monotone. Using the triangle inequality we obtain $|(\alpha(A_i)+\varepsilon)| \leq |(\alpha(A_i))|+|(\varepsilon)| = |(\alpha(A_i))| + \varepsilon|(1)|$ and thus $\alpha(A) \leq |(\alpha(A_i))| + \varepsilon[1 + |(1)|]$.

Analogously, if we let $d_i = \max\{\alpha(A_i) - \varepsilon, 0\}$, we have $\alpha(A) + \varepsilon \geq \|(x_n^i) - (x_m^i)\| = |(\|x_n^i - x_m^i\|)| \geq |(d_i)| \geq |(\alpha(A_i))| - |(\varepsilon)| = |(\alpha(A_i))| - \varepsilon|(1)|$. Therefore $\alpha(A) \geq |(\alpha(A_i))| - \varepsilon[1 + |(1)|]$ and because $\varepsilon > 0$ is arbitrary, we have $\alpha(A) = |(\alpha(A_i))|$. This formula is clearly true if some of the A_i are singletons.

On the other hand, using Lemma IX.2.5, Lemma 3.1 and Lemma 3.3, we can assume that, for a given $\delta > 0$, there exists $v_\delta^i \in X_i$, for every $i = 1, \ldots, k$ such that $\chi(A_i) - \delta \leq \|x_n^i - v_\delta^i\| \leq \chi(A_i) + \delta$ for $i = 1, \ldots, k$ and for infinitely many x_n. As a consequence of this, $\|(x_n^i) - (v_\delta^i)\| = |(\|x_n^i - v_\delta^i\|)| \leq \left|\left(\chi(A_i) + \delta\right)\right| \leq \left|\left(\chi(A_i)\right)\right| + |(\delta)| = |(\chi(A_i))| + \delta|(1)|$, which means that $\chi(A) \leq |(\chi(A_i))| + \delta|(1)|$ and, as $\delta > 0$ is arbitrary, we infer that $\chi(A) \leq |(\chi(A_i))|$.

Finally, if $\chi(A) < |(\chi(A_i))|$, choose s such that $\chi(A) < s < |(\chi(A_i))|$. Then there exist $(w^i) \in X_1 \oplus \cdots \oplus X_k$ and infinitely many $(x_n^i) \in A$, such that $s > \|(x_n^i) - (w^i)\| = |(\|x_n^i - w^i\|)|$. However, due to the minimality of A_i $(i = 1, \ldots, k)$, except for a finite number of x_n we have $\|x_n^i - w^i\| \geq \chi(A_i)$ and therefore $|(\|x_n^i - w^i\|)| \geq |(\chi(A_i))| = |(\chi(A_i))| > s$ which is a contradiction. Thus $\chi(A) = |(\chi(A_i))|$ and the same equality is clearly true if some of the A_i are singletons. $\qquad \square$

THEOREM 6.3. *Let $X_1, \ldots, X_k$ be separable Banach spaces, $|\cdot|$ a monotone norm on $\mathbb{R}^k$ and $X_1 \oplus \cdots \oplus X_k$ the direct sum with the induced norm. Then*

$$\delta'(X_1 \oplus \cdots \oplus X_k) = \min_{1 \leq i \leq k} \{\delta'(X_i)\}, \quad \delta(X_1 \oplus \cdots \oplus X_k) = \max_{1 \leq i \leq k} \{\delta(X_i)\}$$

and

$$\gamma(X_1 \oplus \cdots \oplus X_k) = \frac{\max\{\delta(X_1), \ldots, \delta(X_k)\}}{\min\{\delta'(X_1), \ldots, \delta'(X_k)\}}.$$

Proof. Let A be an α-minimal set in $X_1 \oplus \cdots \oplus X_k$. Taking subsequences, if necessary, we can suppose that A has α-minimal and χ-minimal or singleton projections A_i $(i = 1, \ldots, k)$. Moreover, using the previous lemma, the monotonicity of the norm and the fact that $\delta'(X_i)\alpha(A_i) \leq \chi(A_i) \leq \delta(X_i)\alpha(A_i)$ for $i = 1, \ldots, k$, we have

$$\frac{\chi(A)}{\alpha(A)} = \frac{|(\chi(A_i))|}{|(\alpha(A_i))|} \leq \frac{|(\delta(X_i)\alpha(A_i))|}{|(\alpha(A_i))|} \leq \max\{\delta(X_1), \ldots, \delta(X_k)\},$$

$$\frac{\chi(A)}{\alpha(A)} = \frac{|(\chi(A_i))|}{|(\alpha(A_i))|} \geq \frac{|(\delta'(X_i)\alpha(A_i))|}{|(\alpha(A_i))|} \geq \min\{\delta'(X_1), \ldots, \delta'(X_k)\},$$

that is,

$$\min_{1 \leq i \leq k} \{\delta'(X_i)\} \leq \delta'(X_1 \oplus \cdots \oplus X_k) \quad \text{and} \quad \delta(X_1 \oplus \cdots \oplus X_k) \leq \max_{1 \leq i \leq k} \{\delta(X_i)\}.$$

On the other hand it is clear that the converse inequalities hold because, given an α-minimal set $A_i = \{x_n^i : n \in \mathbb{N}\}$ in X_i, the set $\{(x_n^i e_i) : n \in \mathbb{N}\}$ (e_i is the i-th canonical vector of $\mathbb{R}^k$) is α-minimal in $X_1 \oplus \cdots \oplus X_k$. $\qquad \square$

Remark 6.4.

(a) The packing rates for $X_1 \oplus \cdots \oplus X_k$ are independent of the particular monotone norm given in $\mathbb{R}^k$.

(b) In general, the packing rate corresponding to $X_1 \oplus \cdots \oplus X_k$ is worse than the one corresponding to X_i, for every $i = 1, \ldots, k$. For instance, we know that $\gamma(\ell^i) = 1$ for $i = 1, \ldots, k$ but $\gamma(\ell^1 \oplus \cdots \oplus \ell^k) = 2^{1-\frac{1}{k}}$ which is close to 2 for a large k.

Now, we shall study a similar problem for the infinite direct sum of separable Banach spaces. Although it can be tacked for a monotone norm in the substitution space, we only consider the case of an ℓ^p-sum $(1 \le p < +\infty)$.

Using the separability of ℓ^p, the following lemma can be easily proved.

LEMMA 6.5. *Let $\{X_k\}$ be a sequence of separable Banach spaces and $1 \le p < +\infty$. Then, the set formed by all elements $x = (x^k)$ such that $x^k \in X_k$ and $\sum_{k=1}^{+\infty} \|x^k\|^p < +\infty$ with the norm*

$$\|x\|^p = \sum_{k=1}^{+\infty} \|x^k\|^p$$

is a separable Banach space denoted by $\oplus_p X_k$.

DEFINITION 6.6. *Let $\{X_k\}$ be a sequence of Banach spaces and $1 \le p < +\infty$. The subset A of $\oplus_p X_k$ is said to be an $\alpha\chi$-regular (regular, for short) set if it satisfies the following conditions:*

(1) *A is an α-minimal and χ-minimal set, $A = \{x_n : n \in \mathbb{N}\}$ which also satisfies $\alpha(A) = \lim_{n,m,n\ne m} \|x_n - x_m\|$.*

(2) *For every $k \in \mathbb{N}$, the projection A_k of A into X_k is an α-minimal and χ-minimal set or a singleton.*

(3) *For each $k \in \mathbb{N}$, $\lim\limits_{n,m\; n\ne m} \|x_n^k - x_m^k\| = \alpha(A_k)$ and for every $\varepsilon > 0$, there exists*

$$v_\varepsilon^k \in X_k \text{ such that } \chi(A_k) - \frac{\varepsilon}{2^k} \le \|x_n^k - v_\varepsilon^k\| \le \chi(A_k) + \frac{\varepsilon}{2^k} \text{ for all } n \ge k.$$

Using the separability of X_k $(k \in \mathbb{N})$ and $\oplus_p X_k$, a diagonal argument easily proves the following lemma.

LEMMA 6.7. *Let $\{X_k\}$ be a sequence of separable Banach spaces, and $1 \le p < +\infty$. For every α-minimal set A of $\oplus_p X_k$ there exists an infinite subset A_0 of A such that A_0 is a regular set and $\chi(A_0) = \chi(A)$.*

LEMMA 6.8. *Let A be a regular set in $\oplus_p X_k$ and A_k the projection of A on X_k for every $k \in \mathbb{N}$. Then, $\sum_{k=1}^{+\infty} \alpha^p(A_k)$ and $\sum_{k=1}^{+\infty} \chi^p(A_k)$ are convergent series.*

Proof. We shall prove that $\sum_{k=1}^{+\infty} \alpha^p(A_k) \leq \alpha^p(A)$. Otherwise there exists $k_0 \in \mathbb{N}$ such that $\sum_{k=1}^{k_0} \alpha^p(A_k) > \alpha^p(A)$. For an arbitrary positive number ε, we can take n and m large enough so that for $k = 1, 2, \ldots, k_0$ we have

$$\alpha^p(A_k) - \frac{\varepsilon}{2^{k+1}} \leq \|x_n^k - x_m^k\|^p \leq \alpha^p(A_k) + \frac{\varepsilon}{2^{k+1}}$$

and therefore

$$\|x_n - x_m\|^p \geq \sum_{k=1}^{k_0} \|x_n^k - x_m^k\|^p \geq \sum_{k=1}^{k_0} \left(\alpha^p(A_k) - \frac{\varepsilon}{2^{k+1}}\right) > \sum_{k=1}^{k_0} \alpha^p(A_k) - \frac{\varepsilon}{2}.$$

Hence $\alpha^p(A) = \lim_{n,m;n\neq m} \|x_n - x_m\|^p \geq \sum_{k=1}^{k_0} \alpha^p(A_k) - \frac{\varepsilon}{2}$ and therefore $\alpha^p(A) \geq \sum_{k=1}^{k_0} \alpha^p(A_k)$ because ε is arbitrary.

The series $\sum_{k=1}^{+\infty} \chi^p(A_k)$ is convergent since the inequality $\chi(B) \leq \alpha(B)$ is true for any bounded subset B of a metric space X. $\quad\square$

LEMMA 6.9. *Let a, b, x be positive numbers, $1 \leq p < +\infty$. Then*

$$(a + x)^p \leq a^p(1 + b)^p + \left(\frac{1}{b} + 1\right)^p x^p.$$

A proof can easily be obtained by considering the two cases $ab \leq x$ and $ab \geq x$.

Next lemma might be summarized as follows: Taking a "good" centre for the balls in each X_k, they form a vector which belongs to the direct sum space, and this vector is a "good" centre for the balls in this space.

LEMMA 6.10. *Let $1 \leq p < +\infty$ and $\{x_n\}$ be a sequence in $\oplus_p X_k$ such that $\lim_n \|x_n - z\|$ exists for any $z \in \oplus_p X_k$, and $\lim_{n\to\infty} \|x_n^k - z^k\| = \phi_k(z^k)$ exists for any $z^k \in X_k$ and every $k \in \mathbb{N}$.*

Let $\alpha_k = \inf\{\phi_k(z^k) : z^k \in X_k\}$ for every $k \in \mathbb{N}$ and given any $\varepsilon > 0$, let $v_\varepsilon^k \in X_k$ be a vector with $\phi_k^p(v_\varepsilon^k) - \alpha_k^p < \dfrac{\varepsilon}{2^{k+1}}$. Then:

(a) $v_\varepsilon = (v_\varepsilon^k)$ *belongs to $\oplus_p X_k$.*

(b) $\displaystyle\lim_{n\to\infty} \|x_n - v_\varepsilon\|^p \leq \lim_{n\to\infty} \|x_n - z\|^p + \varepsilon$ *for any $z \in \oplus_p X_k$.*

Proof. (a) Since $\{x_n\}$ is a bounded sequence, there exists $M > 0$ such that $\|x_n\| \le M$ for every $n \in \mathbb{N}$. If $v_\varepsilon = (v_\varepsilon^k)$ does not belong to $\oplus_p X_k$, we conclude that the series $\sum_{k=1}^\infty \|v_\varepsilon^k\|^p$ is divergent, which means that for any $H > 0$ there exists $k_1 \in \mathbb{N}$ such that $\sum_{k=1}^{k_1} \|v_\varepsilon^k\|^p > H$.

Let us take an H such that $H - 2^p M^p = 2^{p-1}\varepsilon > 0$.

Since α_k is an infimum, we have in particular that $\lim_{n \to \infty} \|x_n^k\|^p \ge \alpha_k^p$ and, according to the definition of v_ε^k, we also have $\alpha_k^p > \lim_{n \to \infty} \|x_n^k - v_\varepsilon^k\|^p - \varepsilon/2^{k+1}$. Hence, for a large enough n and $k = 1, \ldots, k_1$, $\|x_n^k - v_\varepsilon^k\|^p \le \|x_n^k\|^p + \varepsilon/2^k$. If we also use the inequality $(a+b)^p \le 2^{p-1}(a^p + b^p)$ for $a > 0$, $b > 0$, which follows from the convexity of the function $x \to x^p$, we have

$$H < \sum_{k=1}^{k_1} \|v_\varepsilon^k\|^p \le \sum_{k=1}^{k_1} \left(\|v_\varepsilon^k - x_n^k\| + \|x_n^k\|\right)^p$$

$$\le 2^{p-1}\sum_{k=1}^{k_1'} \|x_n^k - v_\varepsilon^k\|^p + 2^{p-1}\sum_{k=1}^{k_1} \|x_n^k\|^p$$

$$\le 2^{p-1}\sum_{k=1}^{k_1} \left(\|x_n^k\|^p + \frac{\varepsilon}{2^k}\right) + 2^{p-1}\sum_{k=1}^{k_1} \|x_n^k\|^p$$

$$= 2^p \sum_{k=1}^{k_1} \|x_n^k\|^p + 2^{p-1}\sum_{k=1}^{k_1} \frac{\varepsilon}{2^k}$$

$$\le 2^p\|x_n\|^p + 2^{p-1}\varepsilon \le 2^p M^p + 2^{p-1}\varepsilon = H$$

which is a contradiction. So, v_ε belongs to $\oplus_p X_k$.

(b) If for any w belonging to $\oplus_p X_k$ we let $\phi(w) = \lim_{n \to \infty} \|x_n - w\|$, then we want to show that, for an arbitrary $\varepsilon > 0$, $\phi^p(v_\varepsilon) \le \phi^p(z) + \varepsilon$ for any $z \in \oplus_p X_k$.

Suppose, by way of contradiction, that there exists $z \in \oplus_p X_k$ such that $\phi^p(z) + \varepsilon < \phi^p(v_\varepsilon)$. Let $0 < \delta < \min\left\{1, \dfrac{\varepsilon}{2(\phi^p(z) + 4)}\right\}$, $b > 0$ be such that $(1+b)^p \le 1 + \delta$ and $\delta' > 0$ be such that $2^{p-1}\left(\dfrac{b+1}{b}\right)^p \delta' < \delta$.

Since $v_\varepsilon \in \oplus_p X_k$ and $z \in \oplus_p X_k$, there exists $k_0 \in \mathbb{N}$ such that $\sum_{k > k_0} \|v_\varepsilon^k\|^p < \delta'/2$ and $\sum_{k > k_0} \|z^k\|^p < \delta'/2$. Moreover, there exists $n_0 \in \mathbb{N}$ such that if $n > n_0$ and $k = 1, \ldots, k_0$, we have

$$\left| \|x_n - v_\varepsilon\|^p - \phi^p(v_\varepsilon) \right| < \delta, \qquad \left| \|x_n - z\|^p - \phi^p(z) \right| < \delta,$$

$$\|x_n^k - v_\varepsilon^k\|^p \le \|x_n^k - z^k\|^p + \frac{\varepsilon}{2^{k+1}}.$$

Thus for $n > n_0$ using Lemma 6.9 we have

$$\phi^p(v_\varepsilon) < \|x_n - v_\varepsilon\|^p + \delta = \sum_{k \leq k_0} \|x_n^k - v_\varepsilon^k\|^p + \sum_{k > k_0} \|x_n^k - v_\varepsilon^k\|^p + \delta$$

$$\leq \sum_{k \leq k_0} \|x_n^k - z^k\|^p + \sum_{k > k_0} \left(\|x_n^k - z^k\| + \|z^k - v_\varepsilon^k\|\right)^p + \delta + \frac{\varepsilon}{2}$$

$$\leq \sum_{k \leq k_0} \|x_n^k - z^k\|^p + (1+b)^p \sum_{k > k_0} \|x_n^k - z^k\|^p$$

$$+ \left(\frac{1+b}{b}\right)^p \sum_{k > k_0} \left(\|z^k\| + \|v_\varepsilon^k\|\right)^p + \frac{\varepsilon}{2} + \delta$$

$$\leq (1+b)^p \sum_{k=1}^{+\infty} \|x_n^k - z^k\|^p + \left(\frac{1+b}{b}\right)^p \sum_{k > k_0} \left(\|z^k\| + \|v_\varepsilon^k\|\right)^p + \delta + \frac{\varepsilon}{2}$$

$$\leq (1+b)^p \|x_n - z\|^p + \left(\frac{1+b}{b}\right)^p 2^{p-1} \sum_{k > k_0} \left(\|z^k\|^p + \|v_\varepsilon^k\|^p\right) + \frac{\varepsilon}{2} + \delta$$

$$\leq (1+b)^p \|x_n - z\|^p + \left(\frac{1+b}{b}\right)^p 2^{p-1} \delta' + \frac{\varepsilon}{2} + \delta$$

$$< (1+\delta)\|x_n - z\|^p + 2\delta + \frac{\varepsilon}{2} < (1+\delta)(\phi^p(z) + \delta) + 2\delta + \frac{\varepsilon}{2}$$

$$< (1+\delta)\phi^p(z) + 4\delta + \frac{\varepsilon}{2} = (\phi^p(z) + 4)\delta + \phi^p(z) + \frac{\varepsilon}{2} < \phi^p(z) + \varepsilon < \phi^p(v_\varepsilon)$$

which is a contradiction.   □

We are now in a position to prove our main theorem.

THEOREM 6.11. *Let $\{X_k\}$ be a sequence of separable Banach spaces and let $1 \leq p < +\infty$. Then*

$$\delta'(\oplus_p X_k) = \inf_{k \in \mathbb{N}} \{2^{-\frac{1}{p}}, \delta'(X_k)\}, \quad \delta(\oplus_p X_k) = \sup_{k \in \mathbb{N}} \{2^{-\frac{1}{p}}, \delta(X_k)\},$$

$$\gamma(\oplus_p X_k) = \frac{\sup_{k \in \mathbb{N}} \{2^{-\frac{1}{p}}, \delta(X_k)\}}{\inf_{k \in \mathbb{N}} \{2^{-\frac{1}{p}}, \delta'(X_k)\}}.$$

Proof. Let $A = \{x_n : n \in \mathbb{N}\}$ be an α-minimal sequence in $\oplus_p X_k$ and, for every $k \in \mathbb{N}$, let A_k be the projection of A into X_k. Taking subsequences, if necessary, we can assume that $\lim_{n,m \; n \neq m} \|x_n - x_m\| = \alpha(A)$ and also that for a given $\varepsilon > 0$ and for every $k \in \mathbb{N}$, there exists $v_\varepsilon^k \in X_k$ such that for a large enough n we have

$$\chi^p(A_k) - \frac{\varepsilon}{2^k} \leq \|x_n^k - v_\varepsilon^k\|^p \leq \chi^p(A_k) + \frac{\varepsilon}{2^k}$$

and also that $v_\varepsilon = (v_\varepsilon^k)$ belongs to $\oplus_p X_k$. Moreover, because $\chi(A)$ is an infimum, $\chi^p(A) - \varepsilon \leq \|x_n - v_\varepsilon\|^p$ and from Lemma 6.10, for a large enough n, $\|x_n - v_\varepsilon\|^p \leq \|x_n - z\|^p + \varepsilon$ for every $z \in \oplus_p X_k$. Hence, for a large enough n, $\|x_n - v_\varepsilon\|^p \leq \chi^p(A) + \varepsilon$.

Since the measures of noncompactness α and χ are invariant by translations, we can assume $v_\varepsilon = 0$.

Using Lemma 6.8, there exists $k_0 \in \mathbb{N}$ such that $\sum_{k>k_0} \chi^p(A_k) < \varepsilon$. If we denote $Y = X_1 \times X_2 \times \cdots \times X_{k_0}$ and $Z = (\oplus_p X_k)_{k>k_0}$, each $x \in \oplus_p X_k$ can be expressed as $x = (y, z)$ with $y \in Y$, $z \in Z$.

Since $Y \times Z$ is a finite product, denoting by A_0 and B the first and the second projections of A respectively, we can assume that A_0 and B are either α-minimal and χ-minimal sets or finite sets. From Lemma 6.2, for finite products, we have

$$\alpha^p(A) = \alpha^p(A_0) + \alpha^p(B), \quad \chi^p(A) = \chi^p(A_0) + \chi^p(B),$$

$$\alpha^p(A_0) = \sum_{k \leq k_0} \alpha^p(A_k) \quad \text{and} \quad \chi^p(A_0) = \sum_{k \leq k_0} \chi^p(A_k).$$

We now continue by studying the values for $\alpha(B)$ and $\chi(B)$.

Let $r = \chi(B)$, let n_1 be a natural number large enough so that $r^p - \varepsilon \leq \|z_{n_1}\|^p \leq r^p + \varepsilon$ and, to simplify notation, denote $z_{n_1} \in B$ by z_1. Since $\|z_1\|^p = \sum_{k>k_0} \|x_1^k\|^p < +\infty$, there exists k_1 in $\mathbb{N}$ with $k_0 < k_1$ such that $\sum_{k>k_1} \|x_1^k\|^p < \varepsilon$ and since $\sum_{k_0+1}^{k_1} \chi^p(A_k) < \varepsilon$, there exists n such that $\sum_{k_0+1}^{k_1} \|x_n^k\|^p < \varepsilon$.

If we write

$$a_1 = \left(x_1^{(k_0+1)}, \ldots, x_1^{k_1}, 0, 0, \ldots\right) ; \ a_n = \left(x_n^{(k_0+1)}, \ldots, x_n^{k_1}, 0, 0, \ldots\right)$$

$$b_1 = \left(0, 0, \ldots, 0, x_1^{(k_1+1)}, x_1^{(k_1+2)}, \ldots\right) ; \ b_n = \left(0, 0, \ldots, x_n^{(k_1+1)}, x_n^{(k_1+2)}, \ldots\right)$$

we have

$$\|z_1 - z_n\|^p = \|a_1 - a_n\|^p + \|b_1 - b_n\|^p ; \quad \|z_i\|^p = \|a_i\|^p + \|b_i\|^p \ (i = 1, n)$$

and also the following inequalities:

$$\|b_1\|^p = \sum_{k>k_1} \|x_1^k\|^p < \varepsilon, \qquad \|a_n\|^p = \sum_{k_0 < k \leq k_1} \|x_n^k\|^p < \varepsilon,$$

$$\|a_1\|^p = \|z_1\|^p - \|b_1\|^p > r^p - 2\varepsilon, \qquad \|b_n\|^p = \|z_n\|^p - \|a_n\|^p > r^p - 2\varepsilon.$$

Therefore

$$\|a_1 - a_n\|^p \geq |\ \|a_1\| - \|a_n\|\ |^p$$
$$\geq \|a_1\|^p - p\|a_n\|.\|a_1\|^{p-1} \geq r^p - o(1),$$
$$\|b_n - b_1\|^p \geq |\ \|b_n\| - \|b_1\|\ |^p$$
$$\geq \|b_n\|^p - p\|b_1\|.\|b_n\|^{p-1} \geq r^p - o(1)$$

where $o(1) \to 0$ when $\varepsilon \to 0$.

Hence

$$\alpha^p(B) + \varepsilon \geq \|z_1 - z_n\|^p$$
$$= \|a_1 - a_n\|^p + \|b_1 - b_n\|^p \geq 2\chi^p(B) - o(1).$$

On the other hand

$$\alpha^p(B) - \varepsilon \leq \|z_1 - z_n\|^p = \|a_1 - a_n\|^p + \|b_1 - b_n\|^p$$
$$\leq (\|a_1\| + \|a_n\|)^p + (\|b_1\| + \|b_n\|)^p$$
$$\leq 2\left((r^p + \varepsilon)^{\frac{1}{p}} + \varepsilon^{\frac{1}{p}}\right)^p = 2\chi^p(B) + o(1)$$

where $o(1) \to 0$ when $\varepsilon \to 0$.

Let $\delta'_k = \delta'(X_k)$ and $\delta_k = \delta(X_k)$. Then:

$$\chi^p(A) = \chi^p(A_0) + \chi^p(B) = \sum_{k \leq k_0} \chi^p(A_k) + \chi^p(B)$$

$$\leq \sum_{k \leq k_0} \delta_k^p \alpha^p(A_k) + 2^{-1}(\alpha^p(B) + o(1))$$

$$\leq \max_{k \leq k_0}\{\delta_k^p, 2^{-1}\} \left[\sum_{k \leq k_0} \alpha^p(A_k) + \alpha^p(B) + o(1)\right]$$

$$= \max_{k \leq k_0}\{\delta_k^p, 2^{-1}\}(\alpha^p(A) + o(1)) \leq \sup_{k \in \mathbb{N}}\{\delta_k^p, 2^{-1}\}(\alpha^p(A) + o(1)).$$

while

$$\chi^p(A) = \chi^p(A_0) + \chi^p(B) = \sum_{k \leq k_0} \chi^p(A_k) + \chi^p(B)$$

$$\geq \sum_{k \leq k_0} \delta_k'^p \alpha^p(A_k) + 2^{-1}(\alpha^p(B) - o(1))$$

$$\geq \min_{k \leq k_0}\{\delta_k'^p, 2^{-1}\} \left[\sum_{k \leq k_0} \alpha^p(A_k) + \alpha^p(B) - o(1)\right]$$

$$= \min_{k \leq k_0}\{\delta_k'^p, 2^{-1}\}(\alpha^p(A) - o(1)) \geq \inf_{k \in \mathbb{N}}\{\delta_k'^p, 2^{-1}\}(\alpha^p(A) - o(1)).$$

Letting $\varepsilon \to 0$, we obtain

$$\inf_{k \in \mathbb{N}}\{\delta_k'^p, 2^{-1}\}\alpha^p(A) \leq \chi^p(A) \leq \sup_{k \in \mathbb{N}}\{\delta_k^p, 2^{-1}\}\alpha^p(A),$$

that is,

$$\inf_{k \in \mathbb{N}}\{\delta_k', 2^{-\frac{1}{p}}\} \leq \frac{\chi(A)}{\alpha(A)} \leq \sup_{k \in \mathbb{N}}\{\delta_k, 2^{-\frac{1}{p}}\},$$

and therefore

$$\delta'(\oplus_p X_k) \geq \inf_{k \in \mathbb{N}} \{2^{-\frac{1}{p}}, \delta'(X_k)\}, \quad \delta(\oplus_p X_k) \leq \sup_{k \in \mathbb{N}} \{2^{-\frac{1}{p}}, \delta(X_k)\}.$$

Finally we shall show that these bounds are attained:

If, for any fixed $k \in \mathbb{N}$, we consider an α-minimal sequence $\{x_n\}$ in X_k, then the sequence $\{\bar{x}_n\}$ defined as $\bar{x}_n^k = x_n$ and $\bar{x}_n^j = 0$ if $j \neq k$ is α-minimal in $\oplus_p X_k$ and

$$\frac{\chi(\{\bar{x}_n\})}{\alpha(\{\bar{x}_n\})} = \frac{\chi(\{\bar{x}_n^k\})}{\alpha(\{\bar{x}_n^k\})}.$$

Thus $\delta'(\oplus X_k) \leq \delta'(X_k)$ and $\delta(\oplus_p X_k) \geq \delta(X_k)$. Since k is arbitrary we obtain

$$\delta'(\oplus X_k) \leq \inf\{\delta'(X_k) : k \in \mathbb{N}\} \quad \text{and} \quad \delta(\oplus_p X_k) \geq \sup\{\delta(X_k) : k \in \mathbb{N}\}.$$

On the other hand if, for every $k \in \mathbb{N}$ we consider a unitary vector $x^k \in X_k$, then the sequence $\{\bar{x}_n\}$ in $\oplus_p X_k$ defined as $\bar{x}_n^k = \delta_{nk} x^k$ is α-minimal and satisfies $\chi(\{\bar{x}_n\}) = 1$ and $\alpha(\{\bar{x}_n\}) = 2^{\frac{1}{p}}$. Thus

$$\delta'(\oplus_p X_k) \leq 2^{-\frac{1}{p}} \leq \delta(\oplus_p X_k).$$

Hence

$$\gamma(\oplus_p X_k) = \frac{\sup_{k \in \mathbb{N}} \{2^{-\frac{1}{p}}, \delta(X_k)\}}{\inf_{k \in \mathbb{N}} \{2^{-\frac{1}{p}}, \delta'(X_k)\}}.$$

$\square$

Remark 6.12. It is clear that $\gamma(\oplus_p X_k) \geq \sup_{k \in \mathbb{N}}\{\gamma(X_k), \gamma(\ell^p)\}$. Moreover, though $\gamma(X_k) = 1$ for every $k \in \mathbb{N}$, $\gamma(\oplus_p X_k) = 2$ is still possible. This is the case if we consider $X_k = \ell^k$ ($k \in \mathbb{N}$). In fact, $\gamma(\oplus_p X_k) = 1$ if and only if $\delta'(X_k) = \delta(X_k) = 2^{-\frac{1}{p}}$ for every $k \in \mathbb{N}$.

REFERENCES

[A]   J. ARIAS DE REYNA, *On r-separated sets in normed spaces*, Proc. Amer. Math. Soc. **112 (4)** (1991), 1087–1094.

[Ab]   A. AMBROSETTI, *Un teorema di esistenza per le equazioni differenziali negli spazi di Banach*, Rend. Sem. Mat. Padova **39** (1967), 349–361.

[AD]   J. ARIAS DE REYNA AND T. DOMÍNGUEZ BENAVIDES, *On a measure of noncompactness in Banach spaces with Schauder basis*, Bolletino U.M.I. **(7) 7-A** (1993), 77–86.

[AD1]   J.M. AYERBE AND T. DOMÍNGUEZ BENAVIDES, *Set-contractions and ball-contractions in L^p-spaces*, J. Math. Anal. Appl. **159 (2)** (1991), 500–506.

[AD2]   J.M. AYERBE AND T. DOMÍNGUEZ BENAVIDES, *Connections between some Banach space coefficients concerning normal structure*, J. Math. Anal. Appl. **172** (1993), 53–61.

[ADF1]   J.M. AYERBE, T. DOMÍNGUEZ BENAVIDES AND S. FRANCISCO CUTILLAS, *Some noncompact convexity moduli for the property (β) of Rolewicz*, Comm. Appl. Nonlinear Anal. **1 (1)** (1994), 87–98.

[ADF2]   J.M. AYERBE, T. DOMÍNGUEZ BENAVIDES AND S. FRANCISCO CUTILLAS, *A modulus for the property (β) of Rolewicz*, Colloq. Math. **73 (2)** (1997), 183–191.

[ADL]   J.M. AYERBE, T. DOMÍNGUEZ BENAVIDES AND G. LÓPEZ ACEDO, *Connections between some measures of noncompactness and associated operators*, Extracta Math. **5 (2)** (1990), 62–64.

[AK]   A.G. AKSOY AND M.A. KHAMSI, *Nonstandard Methods in Fixed Point Theory*, Springer-Verlag, 1990.

[AKPRS] R.R. AKHMEROV, M.I. KAMENSKIĬ, A.S. POTAPOV, A.E. RODKINA AND B.N. SADOVSKIĬ, *Measures of Noncompactness and Condensing Operators*, Birkhäuser Verlag, 1992.

[Al]   D.E. ALSPACH, *A fixed point free nonexpansive map*, Proc. Amer. Math. Soc. **82** (1981), 423–424.

[Am]   D. AMIR, *On Jung's constant and related constants in normed linear spaces*, Pacific J. Math. **118 (1)** (1985), 1–15.

[Ao]   D. ABBOT, *The Biographical Dictionary of Scientists: Mathematicians*, Muller, Blond and White Ltd., 1985.

[As]   E. ASPLUND, *Positivity of duality mappings*, Bull. Amer. Math. Soc. **73** (1967), 200–203.

[AX]   J.M. AYERBE AND H.K. XU, *On certain geometric coefficients of Banach spaces relating to fixed point theory*, Panamerican J. Math. **3 (3)** (1993), 47–59.

[B]   B. BEAUZAMY, *Introduction to Banach Spaces and Their Geometry*, North-Holland, 1986.

[Ba]   J. BARWISE, *Handbook of Mathematical Logic and Foundations of Mathematics*, vol. 90, North-Holland, 1977.

[Ba1]   J. BANAŚ, *On modulus of noncompact convexity and its properties*, Canad. Math. Bull. **30 (2)** (1987), 186–192.

[Ba2]   J. BANAŚ, *Compactness conditions in the geometric theory of Banach spaces*, Nonlinear Anal. **16 (7/8)** (1991), 669–682.

[BG]   J. BANAŚ AND K. GOEBEL, *Measures of Noncompactness in Banach Spaces*, Marcel Dekker, Inc., 1980.

[BKS]   L.P. BELLUCE, W.A. KIRK AND E.F. STEINER, *Normal structure in Banach spaces*, Pacific J. Math. **26** (1968), 433–440.

[BM]   M.S. BRODSKIĬ AND D.P. MILMAN, *On the center of a convex set*, Dokl. Akad. Nau. S.S.S.R. **59** (1948), 837–840 (Russian).

[Bn]   S. BANACH, *Sur les opérations dans les ensembles abstraits et leurs applications*, Fund. Math. **3** (1922), 133–181.

[Bo]   P. BOHL, *Über die Bewegung eines mechanisches Systems in der Nähe einer Gleich-gewichtslage*, J. Reine Angew. Math. **127** (1904), 279–286.

[BoS]   J.M. BORWEIN AND B. SIMS, *Nonexpansive mappings on Banach lattices and related topics*, Houston J. Math. **10** (1984), 339–356.

[BP]   F.E. BROWDER AND W.V. PETRYSHYN, *The solution by iteration of nonlinear functional equations in Banach spaces*, Bull. Amer. Math. Soc. **72** (1966), 571–576.

[Br]   L.E.J. BROUWER, *Über Abbildungen von Mannigfaltigkeiten*, Math. Ann. **71** (1912), 97–115.

[Br1]   F.E. BROWDER, *Fixed point theorems for noncompact mappings in Hilbert spaces*, Proc. Nat. Acad. Sci. USA **43** (1965), 1272–1276.

[Br2]   F.E. BROWDER, *Nonexpansive nonlinear operators in a Banach space*, Proc. Nat. Acad. Sci. USA **54** (1965), 1041–1044.

[Br3]   F.E. BROWDER, *Convergence theorems for sequences of nonlinear operators in Banach spaces*, Math. Z. **100** (1967), 201–225.

[Br4]   F.E. BROWDER, *Nonlinear operators and nonlinear equations of evolution in Banach spaces*, American Mathematical Society, Providence, RI, (Proc. Sympos. Pure Math. Vol. XVIII part 2, 1976.

[BS1]   J. BERNAL AND F. SULLIVAN, *Multidimensional volumes, super-reflexivity and normal structure in Banach spaces*, Illinois J. Math. **27** (1983), 501–503.

[BS2]   J. BERNAL AND F. SULLIVAN, *Banach spaces that have normal structure and are iso-morfic to a Hilbert space*, Proc. Amer. Math. Soc. **90** (1984), 550–554.

[Bw]   R.F. BROWN, *A Topological Introduction to Nonlinear Analysis*, Birkhäuser Verlag, 1993.

[By1]   W.L. BYNUM, *A class of spaces lacking normal structure*, Compositio Math. **25** (1972), 233–236.

[By2]   W.L. BYNUM, *Normal structure of Banach spaces*, Manuscripta Math. **11** (1974), 203–209.

[By3]   W.L. BYNUM, *Normal structure coefficients for Banach spaces*, Pacific J. Math. **86** (1980), 427–436.

[C]   J.A. CLARKSON, *Uniformly convex spaces*, Trans. Amer. Math. Soc. **40** (1936), 396–414.

[Ca]   E. CASINI, *Degree of convexity and product spaces*, Comment. Math. Univ. Carolin. **31** (1990), 637–641.

[CL]   E. CODDINGTON AND N. LEVINSON, *Theory of Ordinary Differential Equations*, McGraw-Hill, 1955.

[CM]   E. CASINI AND E. MALUTA, *Fixed points of uniformly Lipschitzian mappings in spaces with uniformly normal structure*, Nonlinear Anal. **9** (1985), 103–108.

[Co]   C. CORDUNEANU, *Equazioni differenziali negli spazi de Banach. Teoremi di esistenza e prolongabilita*, Atti Accad. Naz. Lincei Rend. Cl. Sci. Fis. Mat. Nat. **XXIII** (1957), 226–230.

[D]   G. DARBO, *Punti uniti in transformazioni a codomio non compatto*, Rend. Sem. Mat. Uni. Padova **24** (1955), 84–92.

[D1]   K. DEIMLING, *Nonlinear Functional Analysis*, Springer-Verlag, 1985.

[D2]   K. DEIMLING, *Periodic solutions of differential equations in Banach spaces*, Manuscripta Math. **24** (1978), 31–44.

[Da1]   M.M. DAY, *Normed Linear Spaces*, Springer-Verlag, 1973.

[Da2]   M.M. DAY, *Uniform convexity in factor and conjugate spaces*, Ann. of Math. **45 (2)** (1944), 375–385.

[DaS]   T. DALBY AND B. SIMS, *Duality map characterizations for Opial conditions*, Bull. Austral. Math. Soc. **53** (1996), 413–417.

[DB]   D. VAN DULTS AND B. SIMS, *Fixed Points of Nonexpansive Mappings and Chebyshev Centres in Banach Spaces with Norms of Type KK*, LNM 991, Springer-Verlag, 1983.

[De]   J. DIEUDONNÉ, *History of Functional Analysis*, North-Holland, 1981.

[DG]   J. DUGUNDGI AND A. GRANAS, *Fixed Point Theory*, PWN-Polish Scientific Pub., 1982.

[Di]   J. DIESTEL, *Sequences and Series in Banach Spaces*, Springer-Verlag, 1984.

[DJS]   M.M. DAY, R.C. JAMES AND S. SWAMINATHAN, *Normed linear spaces which are uniformly convex in every direction*, Can. J. Math. **23 (6)** (1971), 1051–1059.

[DL1]   T. DOMÍNGUEZ BENAVIDES AND G. LÓPEZ ACEDO, *Fixed points of asymptotically contractive mappings*, J. Math. Anal. Appl. **164 (2)** (1992), 447–452.

[DL2]   T. DOMÍNGUEZ BENAVIDES AND G. LÓPEZ ACEDO, *Lower bounds for normal structure coefficients*, Proc. Royal Soc. Edinburgh 121A (1992), 245–252.

[DLX1]   T. DOMÍNGUEZ BENAVIDES, G. LÓPEZ ACEDO AND H.K. XU, *Weak uniform normal structure and iterative fixed points of nonexpansive mappings*, Colloq. Math. **LXVIII** (1995), 17–23.

[DLX2]   T. DOMÍNGUEZ BENAVIDES, G. LÓPEZ ACEDO AND H.K. XU, *On quantitative and qualitative properties of the space ℓ^{pq}*, Houston J. Math. **22 (1)** (1996), 89–99.

[Do1]   T. DOMÍNGUEZ BENAVIDES, *Some properties of the set and ball measures of noncompactness and applications*, J. London Math. Soc. **34 (2)** (1986), 120–128.

[Do2]   T. DOMÍNGUEZ BENAVIDES, *Set-contractions and ball-contractions in some classes of spaces*, J. Math. Anal. Appl. **136 (1)** (1988), 131–140.

[Do3]   T. DOMÍNGUEZ BENAVIDES, *Some topological properties of the 1-set-contractions*, Proc. Amer. Math. Soc. **93 (2)** (1985), 252–254.

[Do4]   T. DOMÍNGUEZ BENAVIDES, *Normal structure coefficients in L_p-spaces*, Proc. Royal Soc. Edinburgh **117A** (1991), 299–303.

[Do5]   T. DOMÍNGUEZ BENAVIDES, *Weak uniform normal structure coefficients in direct sums*, Studia Math. **103 (3)** (1992), 283–290.

[Do6]   T. DOMÍNGUEZ BENAVIDES, *Fixed point theorems for uniformly lipschitzian mappings and asymptotically regular mappings*, Nonlinear Analysis T.M.A. (to appear).

[Do7]   T. DOMÍNGUEZ BENAVIDES, *Modulus of nearly uniform smoothness and Lindenstrauss formulae*, Glasgow J. Math. **37** (1995), 145–153.

[Do8]   T. DOMÍNGUEZ BENAVIDES, *Stability of the fixed point property for nonexpansive mappings*, Houston J. Math. **22 (4)** (1996), 835–849.

[DR1]   T. DOMÍNGUEZ BENAVIDES AND R.J. RODRIGUEZ, *Some geometric coefficients in Orlicz sequence spaces*, Nonlinear Anal. **20 (4)** (1993), 349–358.

[DR2]   T. DOMÍNGUEZ BENAVIDES AND R.J. RODRIGUEZ, *Packing coefficients in direct sum spaces*, Bolletino U.M.I. **7 (9A)** (1995), 377–390.

[DS]   N. DUNFORD AND J.T. SCHWARTZ, *Linear Operators, (3 Vol)*, Interscience, 1971.

[DT]   D.J. DOWNING AND B. TURETT, *Some properties of the characteristic of convexity relating to fixed point theory*, Pacific J. Math. **104** (1983), 343–350.

[Du]   D. VAN DULTS, *Equivalent norms and fixed point property for nonexpansive mappings*, J. London Math. Soc. **25** (1982), 139–144.

[DX]   T. DOMÍNGUEZ BENAVIDES AND H.K. XU, *A new geometrical coefficient for Banach spaces and its applications in fixed point theory*, Nonlinear Anal. **25 (3)** (1995), 311–325.

[E]   M. EDELSTEIN, *A theorem on fixed point under isometries*, Amer. Math. Monthly **70 (3)** (1963), 298–300.

[Ed]   EDITORIAL BOARD, *Stanisław Mazur*, Studia Math. **71** (1982), 223–226.

[F]   T. FIGIEL, *On the moduli of convexity and smothness*, Studia Math. **56** (1976), 121–155.

[FLM]   T. FIEGEL, J. LIDENSTRAUSS AND V.D. MILMAN, *The dimension of almost spherical sections of convex bodies*, Acta Math. **139** (1977), 53–94.

[G]   A.N. GODUNOV, *Peano's theorem in Banach spaces*, Functional Anal. App. **9 (1)** (1975), 53–55.

[Ga]   J. GARCÍA-FALSET, *The fixed point property in Banach spaces with NUS property*, preprint.

[Gar]   A.L. GARKAVI, *On the best net and the best section of sets in a normed space*, Izv. Akad. Nauk. SSSR, Ser. Mat. **26 (1)** (1962), 87–106.

[GeS]   R. GEREMIA AND F. SULLIVAN, *Multi-dimensional volumes and moduli of convexity in Banach spaces*, Ann. Mat. Pur. Appl. **127** (1981), 231–251.

[GGM]   I. GOHBERG, L.S. GOL'DENSHTEĬN AND A.S. MARKUS, *Investigation of some properties of bounded linear operators in connection with their q-norms*, Ucen. Zap. Kishinevsk. Un-ta **29** (1957), 29–36.

[GK1]   K. GOEBEL AND W.A. KIRK, *Topics in Metric Fixed Point Theory*, Cambridge University Press, 1990.

[GK2]   K. GOEBEL AND W.A. KIRK, *A fixed point theorem for transformations whose iterates have uniform Lipschitz constant*, Studia Math. **47** (1973), 135–140.

[GL]   J.P. GOSSEZ AND E. LAMI DOZO, *Some geometric properties related to the fixed point theory for nonexpansive mappings*, Pacific J. Math. **40 (3)** (1972), 565–573.

[Go]   D. GÖHDE, *Zum Prinzip der kontraktiven Abbildung*, Math. Nach. **30** (1965), 251–258.

[Go1]   K. GOEBEL, *Convexity of balls and fixed point theorem for mappings with nonexpansive square*, Compositio Math. **22** (1970), 231–251.

[Go2]   K. GOEBEL, *On the structure of the normal invariant sets for nonexpansive mappings*, Annal. Univ. Mariae Curie-Skłodowska. **29** (1975), 70–72.

[GR]   K. GOEBEL AND S. REICH, *Uniform Convexity, Hyperbolic Geometry and Nonexpansive Mappings*, Marcel Dekker, Inc., 1984.

[Gr1]   J. GORNICKI, *A fixed point theorem for asymptotically regular mapping*, Colloq. Math. **LXIV** (1993), 55–57.

[Gr2]   J. GORNICKI, *Fixed point theorems for asymptotically regular mappings in L^p spaces*, Nonlinear Anal. **17** (1991), 153–159.

[GS]   K. GOEBEL AND T. SĘKOWSKI, *The modulus of noncompact convexity*, Annal. Univ. Mariae Curie-Skłodowska **38** (1984), 41–48.

[H]   P. HARTMANN, *Ordinary Differential Equations*, John Wiley and Sons, 1964.

[Ha]   O. HANNER, *On the uniform convexity of L^p and ℓ^p*, Ark. Math. **3** (1956), 239–244.

[Hl]   P.R. HALMOS, *Measure Theory*, Springer-Verlag, 1974.

[Hu]   R. HUFF, *Banach spaces which are nearly uniformly convex*, Rocky Mountain J. Math. **4** (1980), 743–749.

[I1]   V.I. ISTRĂŢESCU, *On a measure of noncompactness*, Bull. Math. Soc. Sci. Math. R.S. Roumanie (N.S.) **16 (64) n.2** (1972), 195–197.

[I2]   V.I. ISTRĂŢESCU, *Fixed Point Theory*, Reidel Pub. Co., 1981.

[I3]   V.I. ISTRĂŢESCU, *Strict Convexity and Complex Strict Convexity*, Marcel Dekker, 1984.

[J]   R.C. JAMES, *Weak compactness and reflexivity*, Israel J. Math. **2** (1964), 101–119.

[JKP]   J. JAWOROWSKI, W.A. KIRK AND S. PARK, *Antipodal Points and Fixed Points*, Lecture Notes Series Number 28 at Seoul National University, 1995.

[JL]   A. JIMÉNEZ AND E. LLORENS, *Stability of the fixed point property for nonexpansive mappings*, Houston J. Math. **17 (4)** (1991), 251–257.

[K]   S. KAKUTANI, *Topological properties of the unit sphere of a Hilbert space*, Proc. Imp. Acad. Tokyo **14** (1943), 242–245.

[Ka1]   L.A. KARLOVITZ, *Existence of fixed points of nonexpansive mappings in a space without normal structure*, Pacific J. Math. **66** (1976), 153–159.

[Ka2]   L.A. KARLOVITZ, *On nonexpansive mappings*, Proc. Amer. Math. Soc. **55** (1976), 321–325.

[Kh1]   M.A. KHAMSI, *Étude de la Propriété du Point Fixe dans les Espaces de Banach et les Espaces Métriques*, Ph. D. Dissertation, University of Paris VI, 1987.

[Kh2]   M.A. KHAMSI, *On the stability of the fixed point property in ℓ_p*, Rev. Colombiana Mat. **28 no. 1** (1994), 1–6.

[Ki1]   W.A. KIRK, *A fixed point theorem for mappings which do not increase distances*, Amer. Math. Monthly **72** (1965), 1004–1006.

[Ki2]   W.A. KIRK, *Fixed Point Theory for Nonexpansive Mappings, LNM 886*, Springer-Verlag, 1981.

[Ki3]   W.A. KIRK, *The modulus of k-rotundity*, Bolletino U.M.I. **(7) 2A** (1988), 195–201.

[Ki4]   W.A. KIRK, *Nonexpansive mappings in product spaces, set-valued mappings, and k-uniform rotundity*, Nonlinear Functional Analysis (F.E. Browder, ed.), Amer. Math. Soc. Proc. Symp. Pure Math. **45 pt. 2** (1986), 51–64.

[Kl]   V. KLEE, *Some topological properties of convex sets*, Trans. Amer. Math. Soc. **178** (1955), 30–45.

[KL1]   D.N. KUTZAROVA AND T. LANDES, *Nearly uniform convexity of infinite direct sums*, Trans. Indiana Univ. Math. J. **41** (1992), 915–926.

[KL2]   D.N. KUTZAROVA AND T. LANDES, *NUC and related properties of finite direct sums*, Bolletino U.M.I. **8 no. 1** (1994), 45–54.

[KMP]   D.N. KUTZAROVA, E. MALUTA AND S. PRUS, *Property (β) implies normal structure of the dual space*, Rend. Circ. Mat. Palermo **41** (1992), 353–368.

[KN]   G. KREISEL AND M.H.A. NEWMAN, *Luitzen Egbertus Jan Brouwer (1881–1966)*, Biographical Memories of Fellows of the Royal Society of London **15** (1969), 39–68.

[KP]   D.N. KUTZAROVA AND P.L. PAPINI, *On a characterization of property (β) and LUR*, Bollettino U.M.I. **(7) 6-A** (1992), 209–214.

[Kr]   M.A. KRANOSEL'SKIĬ, *Topological Methods in the Theory of Nonlinear Integral Equations*, Pergamon Press, 1964.

[Ku]   D.N. KUTZAROVA, *$k - (\beta)$ and k-nearly uniformly convex Banach spaces*, J. Math. Anal. Appl. **162 (2)** (1991), 322–338.

[Ku1]   K. KURATOWSKI, *Sur les espaces complets*, Fund. Math. **15** (1930), 301–309.

[Ku2]   K. KURATOWSKI, *A Half Century of Polish Mathematics*, PWN-Polish Scientific Publishers, 1980.

[Ku3]   K. KURATOWSKI, *Topology I*, Academic Press, 1966.

[Ku4]   K. KURATOWSKI, *Topology II*, Academic Press, 1968.

[Ku5]   K. KURATOWSKI, *Introduction to Set Theory and Topology*, Pergamon Press, 1956.

[KZ]   M.A. KRANOSELSKIĬ AND P.P. ZABREĬKO, *Geometrical Methods of Nonlinear Analysis*, Springer-Verlag, 1984.

[Kz]   R. KAŁUZA, *The Life of Stefan Banach*, Birkhäuser Verlag, 1996.

[L]   E.A. LIFSHITZ, *Fixed point theorems for operators in strongly convex spaces (Russian)*, Voronez. Gos. Univ. Trudy Mat. Fak. **16** (1975), 23–28.

[L1]   T. LANDES, *Permanence properties of normal structure*, Pacific J. Math. **110** (1984), 125–143.

[L2]   T. LANDES, *Normal structure and the sum-property*, Pacific J. Math. **123** (1986), 127–147.

[Li1]   T.C. LIM, *Asymptotic centres and nonexpansive mappings in conjugate Banach spaces*, Pacific J. Math. **90 (1)** (1980), 135–143.

[Li2]   T.C. LIM, *On the normal structure and related coefficients*, Pacific J. Math. **111** (1984), 357–369.

[Ln1]   P.K. LIN, *Unconditional bases and fixed points of nonexpansive mappings*, Pacific J. Math. **116** (1985), 69–76.

[Ln2]   P.K. LIN, *k-uniform rotundity is equivalent to k-uniform convexity*, J. Math. Anal. Appl. **132** (1988), 349–355.

[LO]   A. LASOTA AND C. OLECH, *Zdzislaw Opial — a mathematician (1930-1974)*, Ann. Polon. Math. **51** (1990), 7–13.

[LS]   P.K. LIN AND Y. STERNFELD, *Convex sets with the lipschitz fixed point property are compact*, Proc. Amer. Math. Soc. **93 (4)** (1985), 633–639.

[LT1]   J. LINDENSTRAUSS AND L. TZAFIRI, *Classical Banach Spaces I*, Springer-Verlag, 1977.

[LT2]   J. LINDENSTRAUSS AND L. TZAFIRI, *Classical Banach Spaces II*, Springer-Verlag, 1979.

[LTX]   P.K. LIN, K.K. TAN AND H.K. XU, *Demiclosedness principle and asymptotic behavior for asymptotically nonexpanxive mappings*, Nonlinear Anal. **24** (1995), 929–946.

[M]   R.H. MARTIN, *Nonlinear Operators and Differential Equations in Banach Spaces*, John Wiley and Sons, 1976.

[Ma]   E. MALUTA, *Uniformly normal structure and related coefficients*, Pacific J. Math. **111 (2)** (1984), 357–369.

[MM]   D.P. MILMAN AND V.D. MILMAN, *The geometry of nested families with empty intersection-structure of the unit sphere of a nonreflexive space*, Trans. Amer. Math. Soc. **85 (2)** (1969), 233–243.

[Mo]   V. MONTESINOS, *Drop property equals reflexivity*, Studia. Math. **87** (1987), 93–100.

[MP]   E. MALUTA AND S. PRUS, *Banach spaces which are dual to k-uniformly convex spaces*, J. Math. Anal. Appl. **209** (1997), 479–491.

[Mu]   B. MAUREY, *Points fixes des contractions sur un convex fermé de L^1*, Séminaire d'Analyse Fonctionelle **Exposé n.8** (1980–81).

[N]   R.D. NUSSBAUM, *The radius of the essential spectrum*, Duke Math J. **37** (1970), 473–478.

[No]   G. NORDLANDER, *The modulus of convexity in normed linear spaces*, Ark. Mat. **4** (1960), 15–17.

[NSW]   J.L. NELSON, K.L. SINGH AND J.H.M. WHITFIELD, *Normal structures and nonexpansive mappings in Banach spaces*, Nonlinear Analysis (edited by Th.M. Rassias) World Sci. Publishing Singapore (1987), 433–492.

[O1]   Z. OPIAL, *Weak convergence of the sequence of sucessive approximations for nonexpansive mappings*, Bull. Amer. Math. Soc. **73** (1967), 591–597.

[O2]   Z. OPIAL, *Lecture notes on nonexpansive and monotone mappings in Banach spaces*, Center for Dynamical Systems (Brown University), 1967.

[O3]   Z. OPIAL, *Continuous parameter dependence in linear sistems of differential equations*, J. Diff. Eq. **3** (1967), 88–91.

[P]   G. PEANO, *Démonstration de l'intégrabilité des équations différentielles ordinaires*, Math. Ann. **37** (1890), 182–228.

[Pa]   J.P. PARTINGTON, *On nearly uniformly convex Banach spaces*, Math. Proc. Camb. Phil. Soc. **93** (1983), 127–129.

[Pe]   W.V. PETRYSHYN, *Remarks on condensing and k-set-contractive mappings*, J. Math. Anal. Appl. **39** (1972), 717–741.

[Ph]   R.R. PHELPS, *A representation theorem for bounded convex sets*, Proc. Amer. Math. Soc. **11** (1960), 976–983.

[Pi]   G. PISIER, *The Volume of Convex Bodies and Banach Space Geometry*, Cambridge University Press, 1989.

[Po]   H. POINCARÉ, *Sur les courbes définies par les équations différentielles IV*, J. Math. Pures et Appl. **2** (1886), 151–217.

[Pr1]   S. PRUS, *On Bynum's fixed point theorem*, Atti. Sem. Mat. Fis. Univ. Modena. **38** (1990), 535–545.

[Pr2]   S. PRUS, *Banach spaces with the uniform Opial property*, Nonlinear Anal. **18** (1992), 697–704.

[Pr3]   S. PRUS, *A remark on a theorem of Turett*, Bull. Polish Acad. Sci. Math. **36** (5–6) (1988), 225–227.

[Pr4]   S. PRUS, *Some estimates for the normal structure coefficient in Banach spaces*, Rend. Circ. Mat. Palermo **XL** (1991), 128–135.

[Pr5]   S. PRUS, *Nearly uniformly smooth Banach spaces*, Bolletino U.M.I. **(7) 3 B** (1989), 507–521.

[R]   H. ROBBINS, *Some complements to Brouwer's fixed point theorem*, Israel J. Math. **5** (1967), 225–226.

[Re]   S. REICH, *Products formulas, Nonlinear semigroups and accretive operators*, J. Funct. Anal. **36** (1980), 147–168.

[Ro]   R.J. RODRÍGUEZ ALVAREZ, *Relaciones entre Operadores Asociados a Distintas Medidas de no Compacidad*, Ph. D. Dissertation, University of Sevilla, 1993.

[Ro1]   S. ROLEWICZ, *On drop property*, Studia Math. **85** (1987), 27–35.

[Ro2]   S. ROLEWICZ, *On Δ-uniform convexity and drop property*, Studia Math. **87** (1987), 181–191.

[S]   J. SCHAUDER, *Der Fixpunktsatz in Funktionalräumen*, Studia Math. **2** (1930), 171–180.

[Sa1]   B.N. SADOVSKIĬ, *On a fixed point principle*, Funkt. Anal. **4** (2) (1967), 74–76.

[Sa2]   B.N. SADOVSKIĬ, *Measures of noncompactness and condensing operators (Russian)*, Problemy Mat. Anal. Sloz. Sistem. **2** (1968), 89–119.

[Sc]   L. SCHWARTZ, *Cours d'Analyse*, Hermann, 1967.

[Se1]   T. SĘKOWSKI, *Functional Analysis and Approximation*, Pitagora Editrice Bologna (Edited by P.L. Papini), 1988.

[Se2]   T. SĘKOWSKI, *On normal structure, stability of fixed point property and the modulus of noncompact convexity*, Rend. Sem. Mat. Fis. Univ. Milano **LVI** (1986), 147–153.

[Sh]   A. SHIELDS, *Felix Hausdorff: Grundzüge der Mengenlehre*, The Mathematical Intelligencer **11** (1) (1989), 6–9.

[Si]   E. SILVERMAN, *Definitions of Lebesgue area for surfaces in metric spaces*, Revista Mat. Univ. Parma **2** (1951), 47–76.

[Si1]   I. SINGER, *Bases in Banach Spaces I*, Springer-Verlag, 1970.

[Si2]   I. SINGER, *Bases in Banach Spaces II*, Springer-Verlag, 1981.

[Sim]   B. SIMS, *Notes on fixed point property*, Queen's University Seminar (1982).

[Sm]   D.R. SMART, *Fixed Point Theorems*, Cambridge University Press, 1974.

[Sp]   M. SPIVAK, *Calculus on Manifols*, W.A. Benjamin, Inc., 1966.

[SS]   T. SĘKOWSKI AND A. STACHURA, *Noncompact smoothness and noncompact convexity*, Atti. Sem. Mat. Fis. Univ. Modena **36** (1988), 329–338.

[ST]   M.A. SMITH AND B. TURETT, *Some examples concerning normal and uniform normal structure in Banach spaces*, J. Austral. Math. Soc. **48A** (1990), 223–234.

[Su]   F. SULLIVAN, *A generalization of uniformly rotund Banach spaces*, Can. J. Math. **31** (1979), 628–636.

[Sz]   S. SZUFLA, *Some remarks on ordinary differential equations in Banach spaces*, Bull.
          Acad. Polonaise Sciences, Série des sciences math., astr. et phys. **XVI (10)** (1968),
          795–800.

[T]   D. TINGLEY, *Noncontractive uniformly lipschitzian semigroups in Hilbert spaces*, Proc.
          Amer. Math. Soc. **92** (1984), 355–361.

[W]   H. WEYL, *Obituary: David Hilbert (1862–1943)*, Obit Notices Roy. Soc. London **4**
          (1944), 547–553.

[W1]   J.R.L. WEBB, *On seminorms of operators*, J. London Math. Soc. **7 (2)** (1973), 337–342.

[W2]   J.R.L. WEBB, *On a property of duality mapping and the A-properness of accretive
          operators*, Bull. London Math. Soc. **13** (1981), 235–238.

[WW]   J.H. WELLS AND L.R. WILLIAMS, *Embeddings and Extensions in Analysis*, Springer-
          Verlag, 1975.

[WZ]   J.R.L. WEBB AND W. ZHAO, *On connections between set and ball measures of non-
          compactness*, Bull. London Math. Soc. **22** (1990), 471–477.

[Y1]   X.T. YU, *k-UR implies NUC*, Kexue Tongbao (Chinese) **28** (1983), 1–3.

[Y2]   X.T. YU, *A geometrical aberrant Banach space with uniformly normal structure*, Bull.
          Austral. Math. Soc. **38** (1988), 99–103.

[Z]   E. ZEIDLER, *Nonlinear Functional Analysis and its Applications I*, Springer-Verlag,
          1986.

[Zh]   G. ZHANG, *Weakly convergent sequence coefficient of product space*, Proc. Amer. Math.
          Soc. **117** (1993), 637–643.

[Zh1]   W. ZHAO, *Geometrical Coefficients and Measures of Noncompactness*, Ph. D. Disser-
          tation, University of Glasgow, 1992.

[Zh2]   W. ZHAO, *Remarks on various measures of noncompactness*, J. Math. Anal. App. **174**
          (1993), 290–297.

SUBJECT INDEX

List of Symbols and Notations

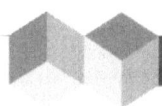

Mathematics with Birkhäuser

BAT · Birkhäuser Advanced Texts / Basler Lehrbücher

M. Rosenblum / J. Rovnyak,
University of Virginia, Charlottesville, VA, USA

Topics in Hardy Classes and Univalent Functions

1994. 264 pages. Hardcover
ISBN 3-7643-5111-X

This book treats classical and contemporary topics in function theory and is accessible after a one-year course in real and complex analysis. It can be used as a text for topics courses or read independently by graduate students and researchers in function theory, operator theory, and applied areas.

The first six chapters supplement the authors' book, Hardy Classes and Operator Theory. The theory of harmonic majorants for subharmonic functions is used to introduce Hardy-Orlicz classes, which are specialized to standard Hardy classes on the unit disk. The theorem of Szegö-Solomentsev characterizes boundary behavior. Half-plane function theory receives equal treatment and features the theorem of Flett and Kuran on existence of harmonic majorants and applications of the Phragmén-Lindelöf principle.

The last three chapters contain an introduction to univalent functions, leading to a self-contained account of Loewner's differential equation and de Branges' proof of the Milin conjecture.

"...I can think of no authors better able than the present ones to give an exposition of the modern theory of univalent functions and they have succeeded admirably...One only hopes that, when it is written, it is as well written as the present volume...The book is written for graduate students. It is a compelling introduction to this fascinating subject and is warmly recommended."

J.M. Anderson, Proceedings of the Edinburgh Mathematical Society, 1995/38

For orders originating from all over the world except USA and Canada:
Birkhäuser Verlag AG
P.O Box 133
CH-4010 Basel/Switzerland
Fax: +41/61/205 07 92
e-mail: farnik@birkhauser.ch

For orders originating in the USA and Canada:
Birkhäuser
333 Meadowland Parkway
USA-Secaurus, NJ 07094-2491
Fax: +1 201 348 4033
e-mail: orders@birkhauser.com

Birkhäuser

Birkhäuser Verlag AG
Basel · Boston · Berlin

VISIT OUR HOMEPAGE **http://www.birkhauser.ch**

81. **H. Upmeier**: Toeplitz Operators and Index Theory in Several Complex Variables, 1996, (ISBN 3-7643-5282-5)

82. **T. Constantinescu**: Schur Parameters, Factorization and Dilation Problems, 1996, (ISBN 3-7643-5285-X)

83. **A.B. Antonevich**: Linear Functional Equations. Operator Approach, 1995, (ISBN 3-7643-2931-9)

84. **L.A. Sakhnovich**: Integral Equations with Difference Kernels on Finite Intervals, 1996, (ISBN 3-7643-5267-1)

85/ **Y.M. Berezansky, G.F. Us, Z.G. Sheftel**: Functional Analysis, Vol. I + Vol. II, 1996,
86. Vol. I (ISBN 3-7643-5344-9), Vol. II (3-7643-5345-7)

87. **I. Gohberg, P. Lancaster, P.N. Shivakumar** (Eds): Recent Developments in Operator Theory and Its Applications. International Conference in Winnipeg, October 2–6, 1994, 1996, (ISBN 3-7643-5414-5)

88. **J. van Neerven** (Ed.): The Asymptotic Behaviour of Semigroups of Linear Operators, 1996, (ISBN 3-7643-5455-0)

89. **Y. Egorov, V. Kondratiev**: On Spectral Theory of Elliptic Operators, 1996, (ISBN 3-7643-5390-2)

90. **A. Böttcher, I. Gohberg** (Eds): Singular Integral Operators and Related Topics. Joint German-Israeli Workshop, Tel Aviv, March 1–10, 1995, 1996, (ISBN 3-7643-5466-6)

91. **A.L. Skubachevskii**: Elliptic Functional Differential Equations and Applications, 1997, (ISBN 3-7643-5404-6)

92. **A.Ya. Shklyar**: Complete Second Order Linear Differential Equations in Hilbert Spaces, 1997, (ISBN 3-7643-5377-5)

93. **Y. Egorov, B.-W. Schulze**: Pseudo-Differential Operators, Singularities, Applications, 1997, (ISBN 3-7643-5484-4)

94. **M.I. Kadets, V.M. Kadets**: Series in Banach Spaces. Conditional and Unconditional Convergence, 1997, (ISBN 3-7643-5401-1)

95. **H. Dym, V. Katsnelson, B. Fritzsche, B. Kirstein** (Eds): Topics in Interpolation Theory, 1997, (ISBN 3-7643-5723-1)

96. **D. Alpay, A. Dijksma, H. de Snoo**: Schur Functions, Operator Colligations, and Reproducing Kernel Pontryagin Spaces, 1997, (ISBN 3-7643-5763-0)

97. **M.L. Gorbachuk / V.I. Gorbachuk**: M.G. Krein's Lectures on Entire Operators, 1997, (ISBN 3-7643-5704-5)

98. **I. Gohberg / Yu. Lyubich** (Eds): New Results in Operator Theory and Its Applications The Israel M. Glazman Memorial Volume, 1997, (ISBN 3-7643-5775-4)

99. **T. Ayerbe Toledano / T. Dominguez Benavides / G. López Acedo**: Measures of Noncompactness in Metric Fixed Point Theory, 1997, (ISBN 3-7643-5794-0)

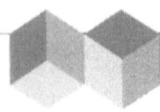
IEOT

Integral Equations and Operator Thoery

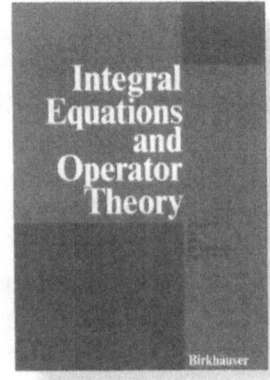

ISSN 0378-620X

IEOT is devoted to the publication of current research in integral equations, operator theory and related topics with emphasis on the linear aspects of the theory.

Aims and Scope

Integral Equations and Operator Theory (IEOT) appears monthly and is devoted to the publication of current research in integral equations, operator theory and related topics with emphasis on the linear aspects of the theory. The journal reports on the full scope of current developments from abstract theory to numerical methods and applications to analysis, physics, mechanics, engineering and others. The journal consists of two sections: a main section consisting of refereed papers and a second consisting of short announcements of important results, open problems, information, etc.

Abstracted/Indexed in:
CompuMath Citation Index, Current Contents,
Mathematical Reviews, Zentralblatt für Mathematik,
Mathematics Abstracts, DB MATH

Subscription Information for 1998
IEOT is published in 3 volumes per year,
and 4 issues per volume
Volumes 30 – 32
approx. 500 pages per volume
Format: 17 x 24 c
Back volumes are available

For orders originating from all over
the world except USA and Canada:
Birkhäuser Verlag AG
P.O Box 133
CH-4010 Basel/Switzerland
Fax: +41/61/205 07 92
e-mail: farnik@birkhauser.ch

For orders originating in the
USA and Canada:
Birkhäuser
333 Meadowland Parkway
USA-Secaurus, NJ 07094-2491
Fax: +1 201 348 4033
e-mail: orders@birkhauser.com

Birkhäuser

Birkhäuser Verlag AG
Basel · Boston · Berlin

VISIT OUR HOMEPAGE **http://www.birkhauser.ch**